My Favorite Fractals
Volume 1
by David E. McAdams

Images in this book were created using Fractal Forge by Uberto Barbini. Fractal Forge can be downloaded from https://sourceforge.net/projects/fractalforge/.

Other Books by David E. McAdams

Parrot Colors – An introduction to the concept of colors. For preschoolers.

Flower Colors – An introduction to the concept of colors. For preschoolers.

Space Colors – An introduction to the concept of colors. For preschoolers.

If I had a Monster – A day in the life of a four-year-old with monsters as daily companions. Coming soon. For preschoolers.

Shapes – An introduction to shapes. For preschoolers.

Numbers – An introduction to the concept of numbers. For grades K-2.

What is Bigger Than Anything? (Infinity) – An introduction to the concept of infinity. For grades 1-3.

Where does the Water Go? – An introduction to waste water treatment for grades 1-4. Coming soon.

Swing sets (Sets) – An introduction to set theory. For grades 2-4.

One Penny, Two – If Sig's penny doubles each day, how long until he can buy a dark green sports car? For grades 3-6.

Learning With Money Activity Kit – Teach large numbers and counting with over $1,000,000 in play money.

My Favorite Fractals – A series of picture books of wondrous fractals presented as high resolution images. For all ages.

The First Million Digits of Pi – The first million digits of pi. For all ages.

e to One Million Digits – The first million digits of the Euler's constant e. For all ages.

The Square Root of 2 to One Million Digits – The first million digits of the square root of 2. For all ages.

Orders of Ten – A book that illustrates orders of ten with dots (1, 10, 100, … dots). For ages 10-15.

Geometric Nets Project Book – 80 geometric nets to copy, cut out, and tape together into 3 dimensional polyhedra. For ages 9 and up.

Geometric Nets Mega Project Book – 253 geometric nets to copy, cut out, and tape together into 3 dimensional polyhedra. For ages 9 and up

For an up to date list, see www.DEMcAdams.com.

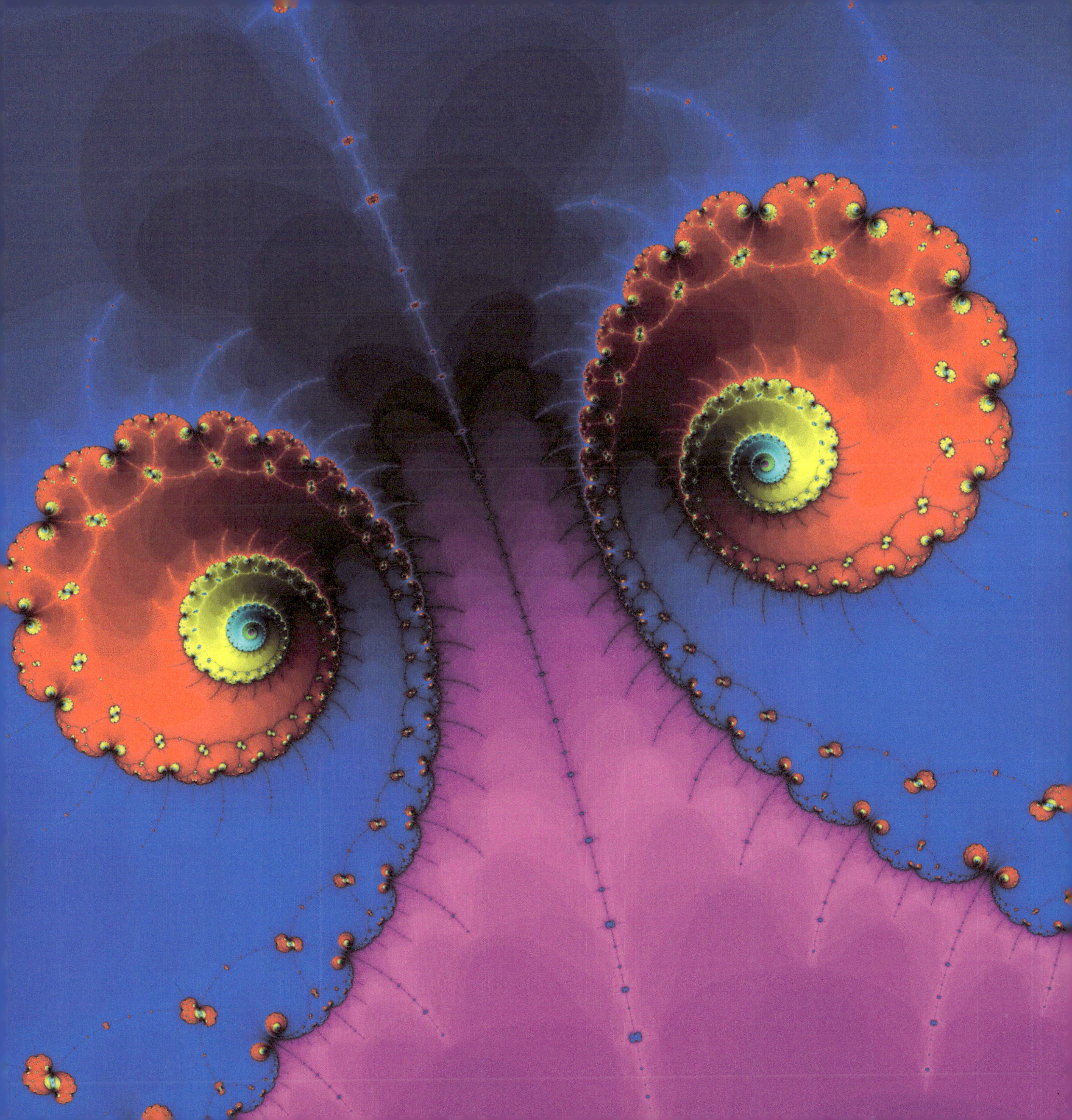

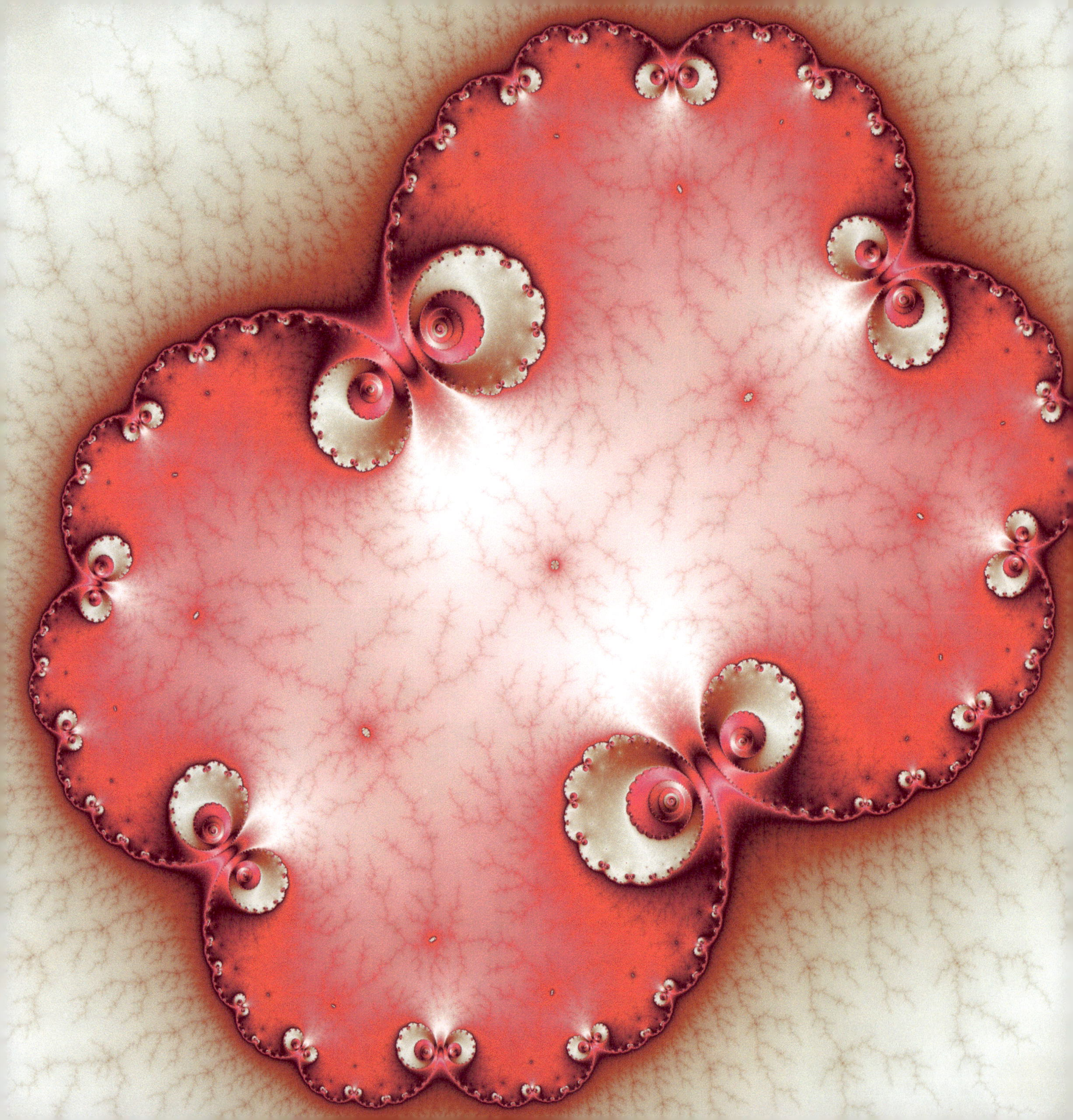

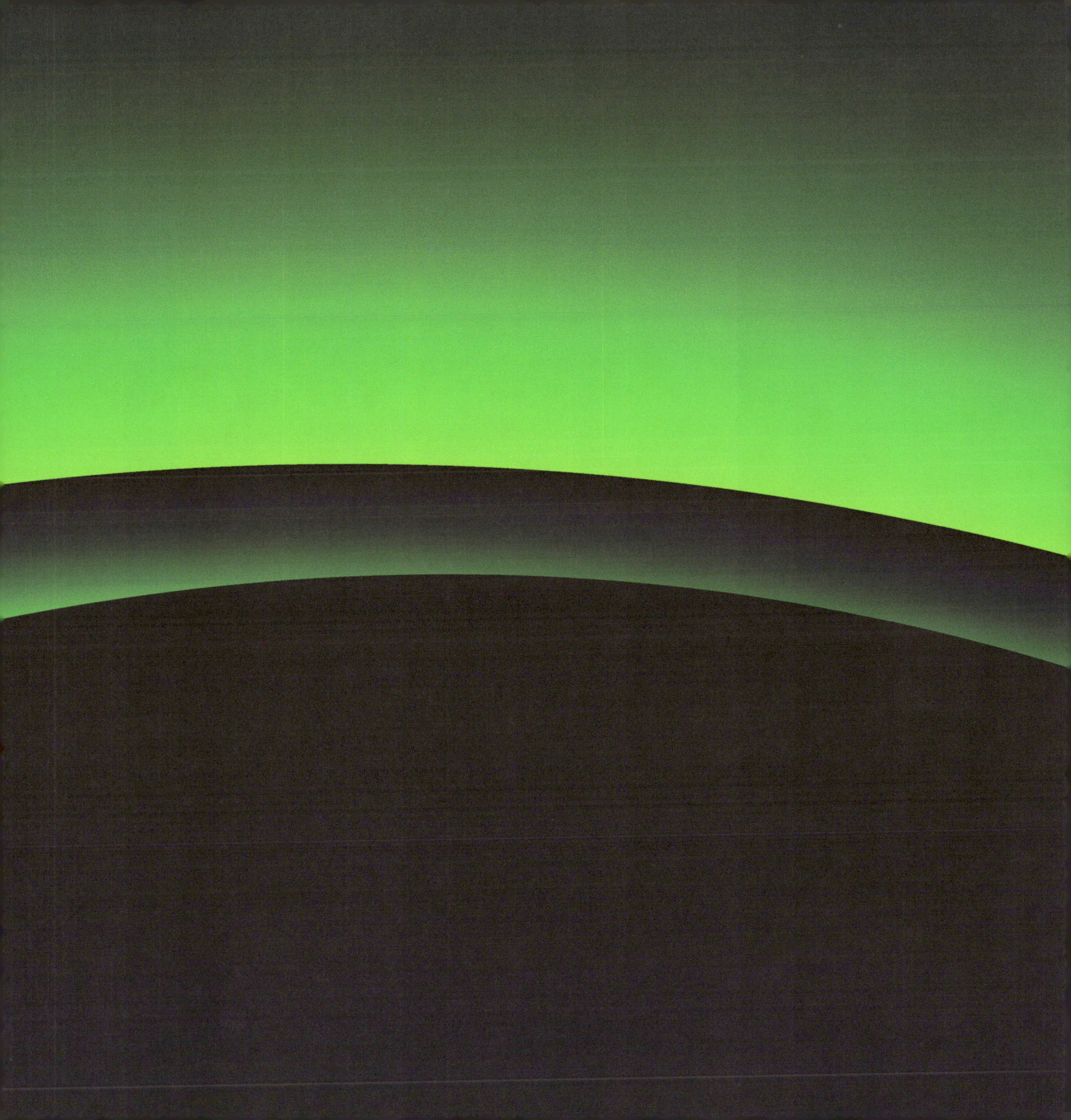

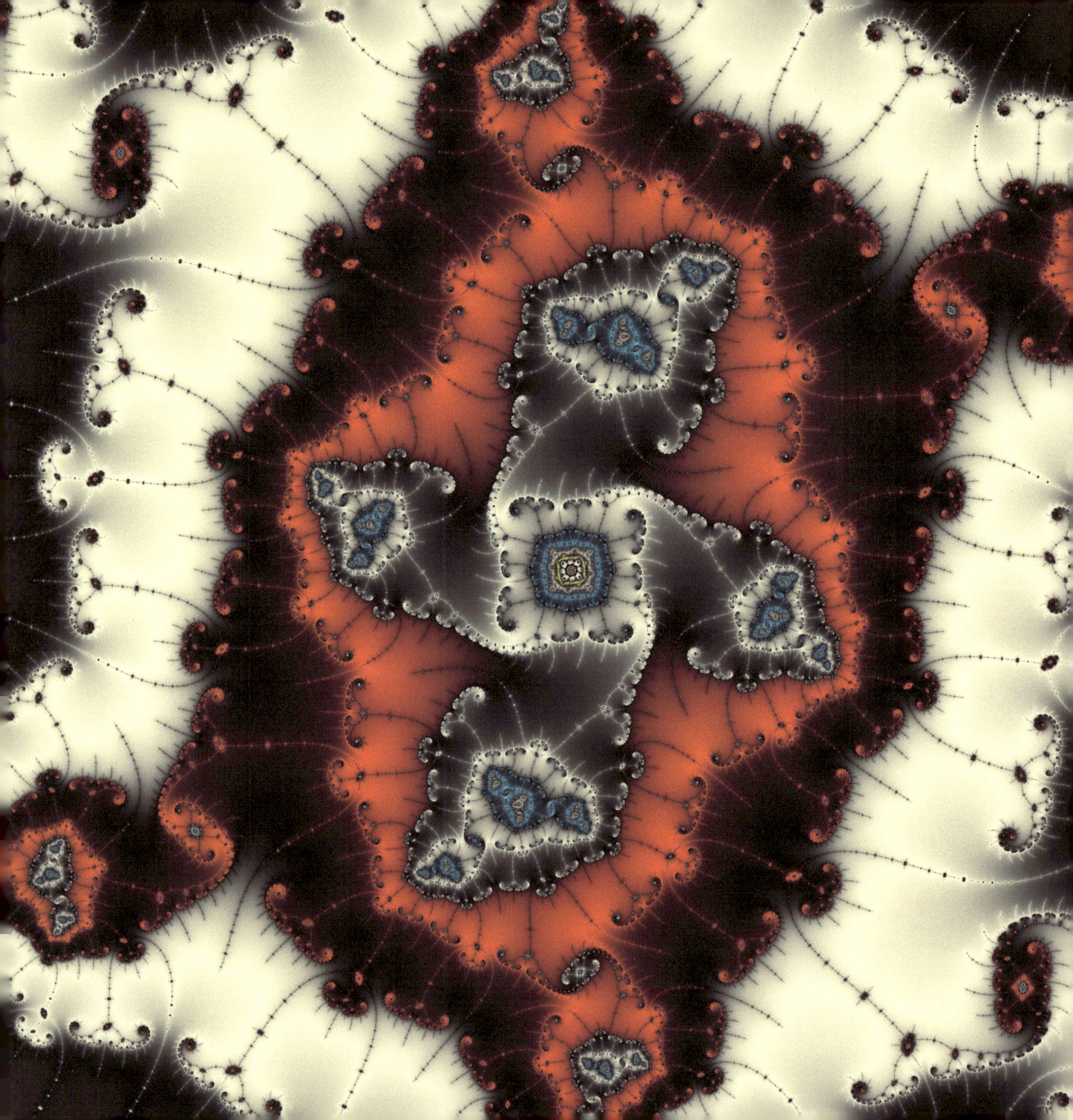

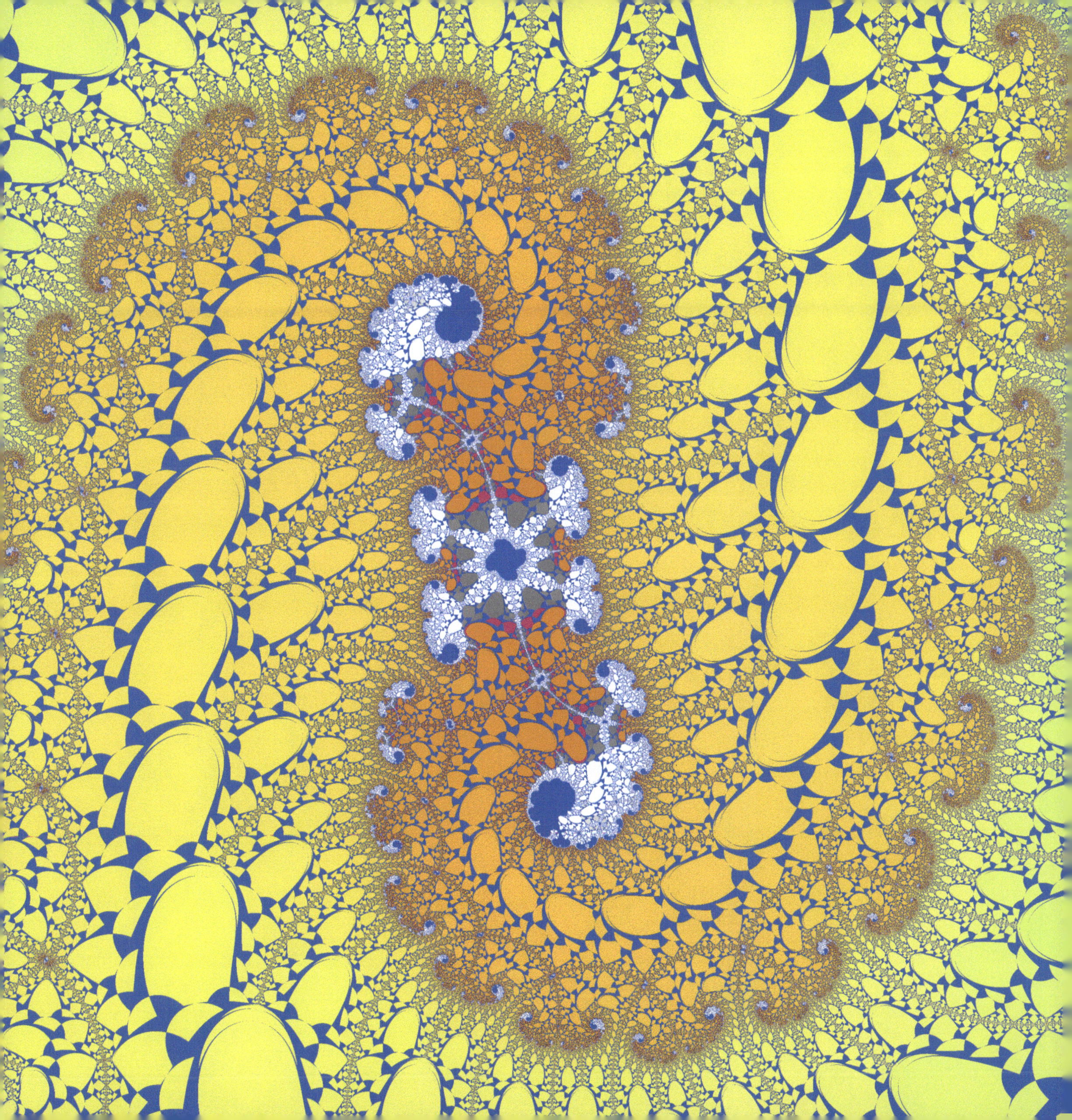

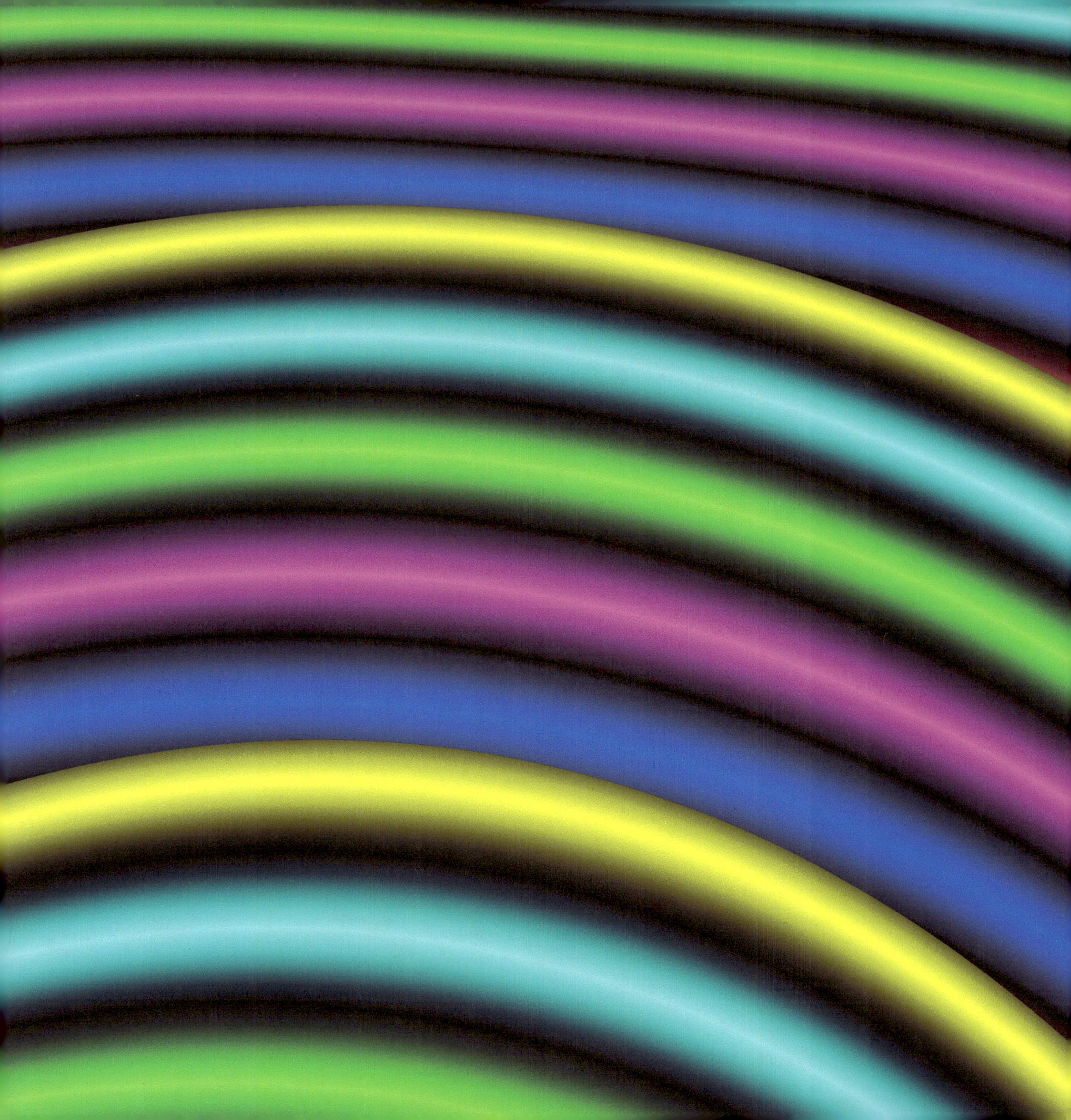

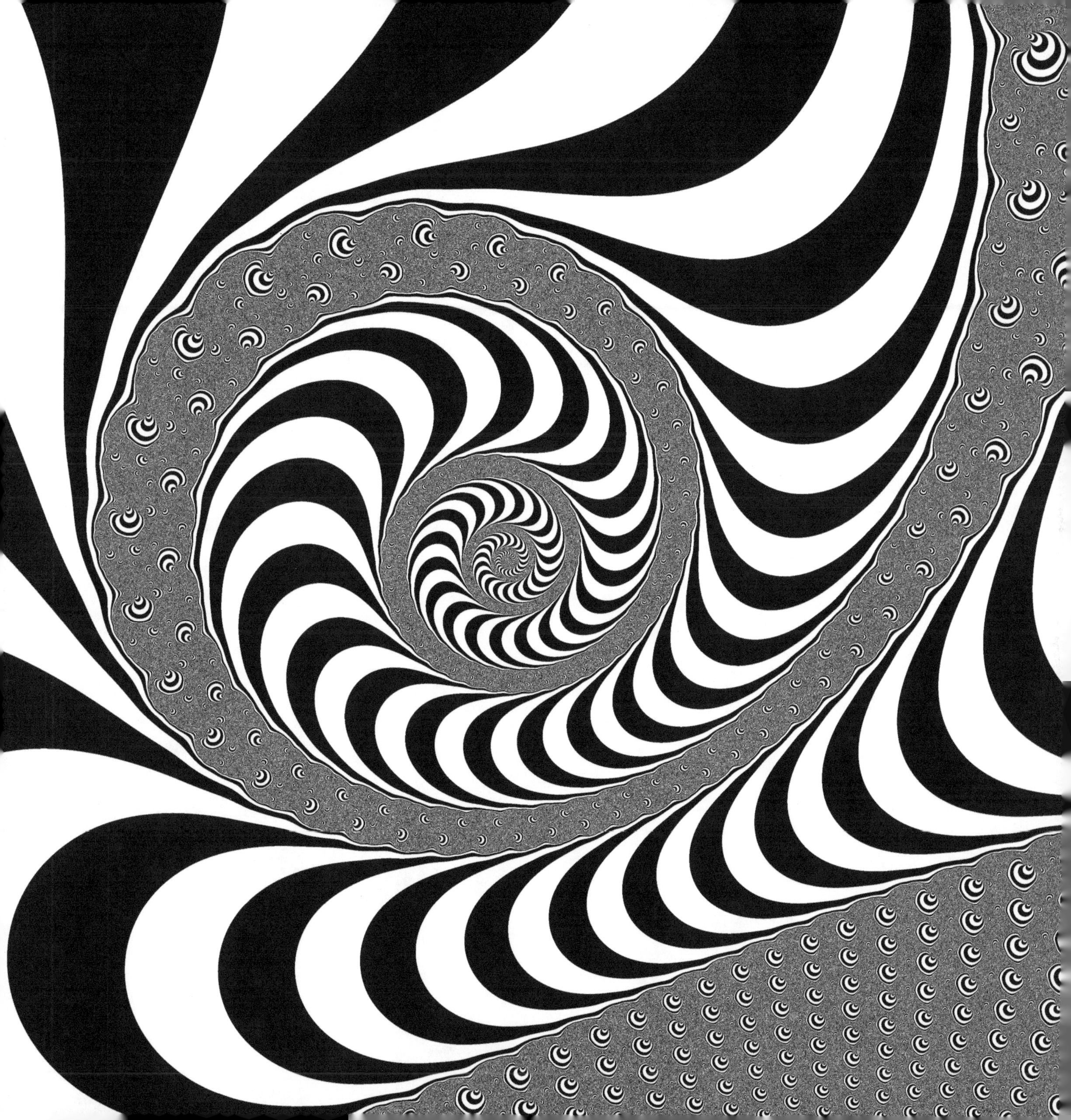

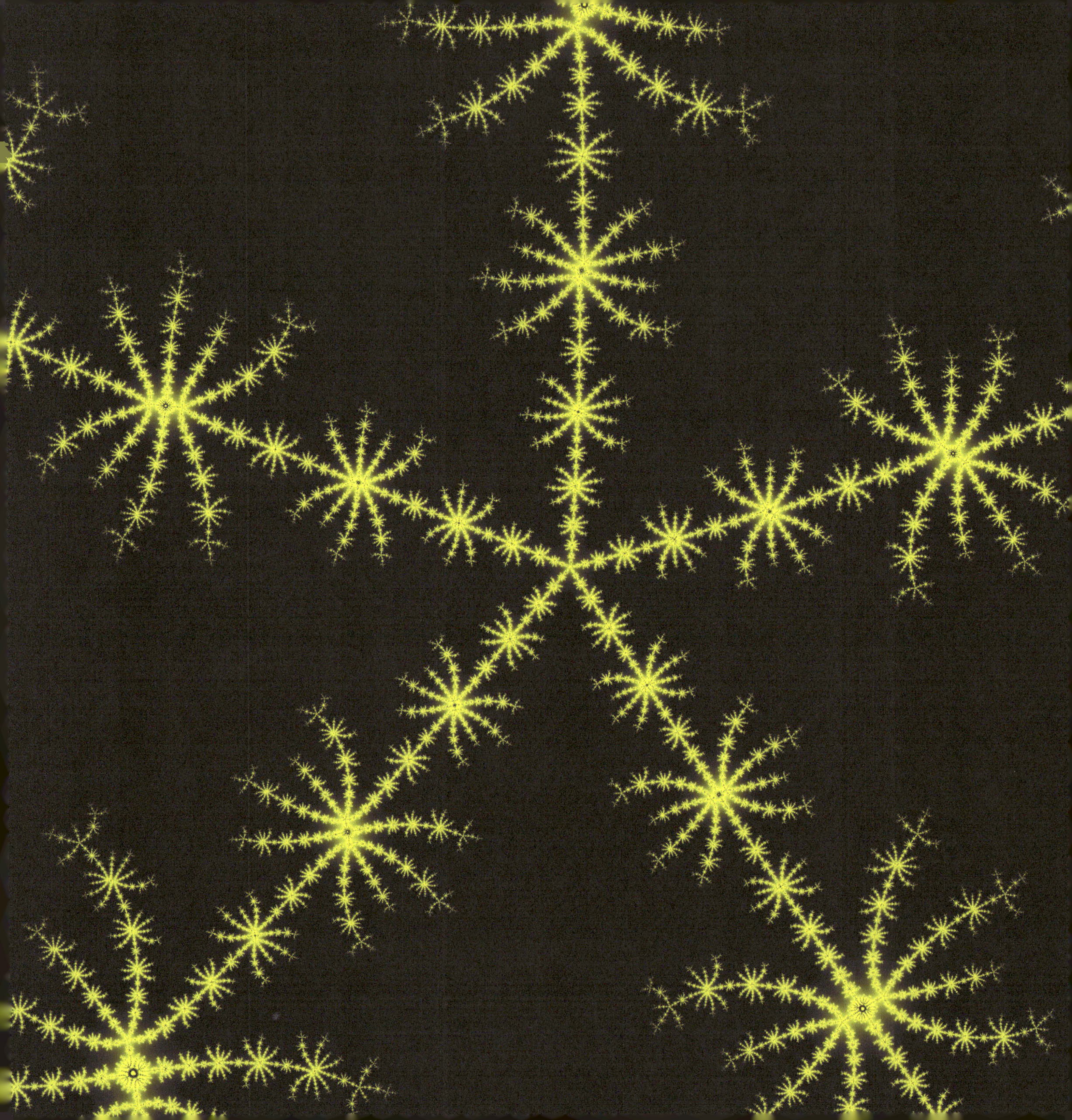

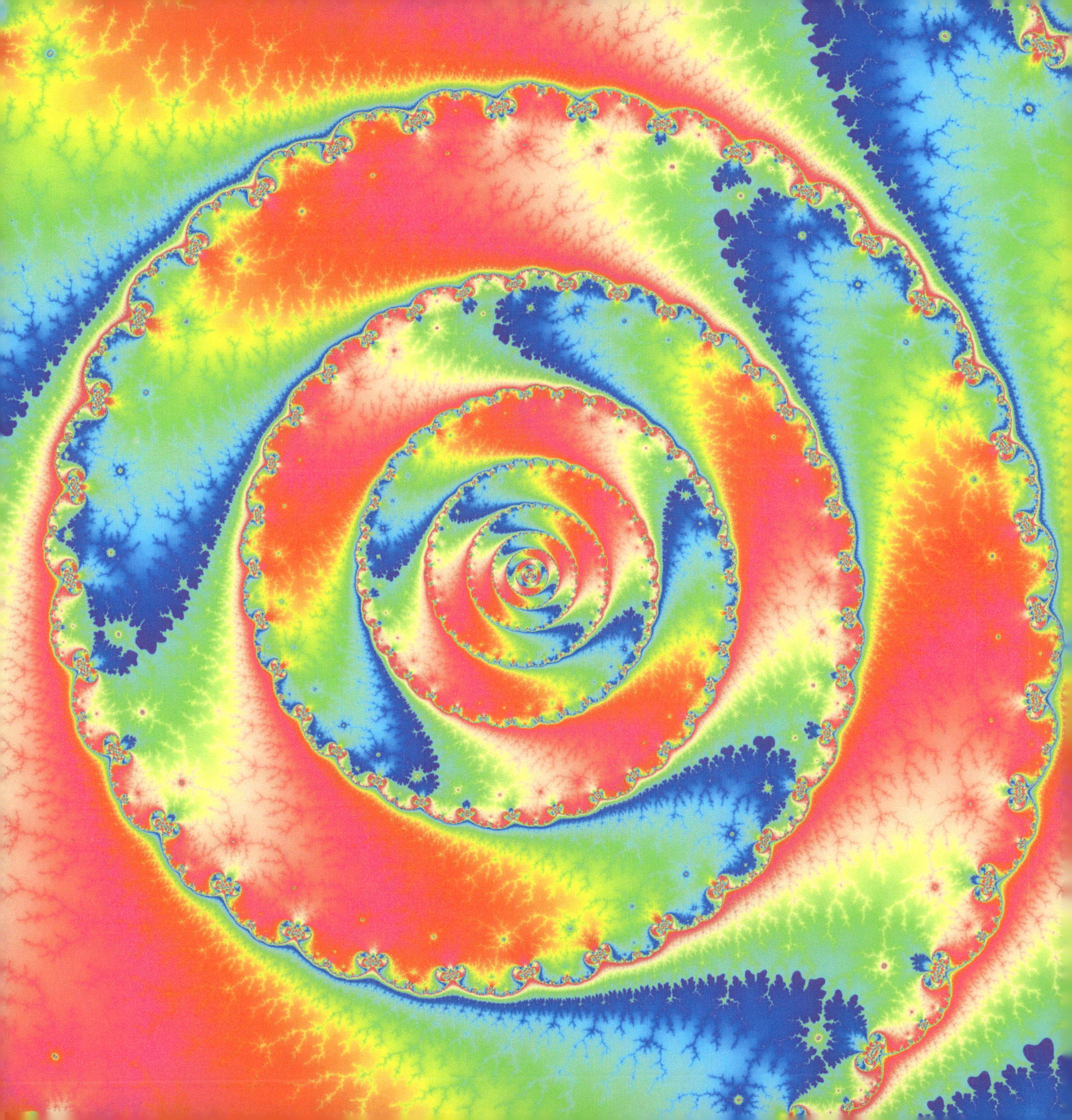

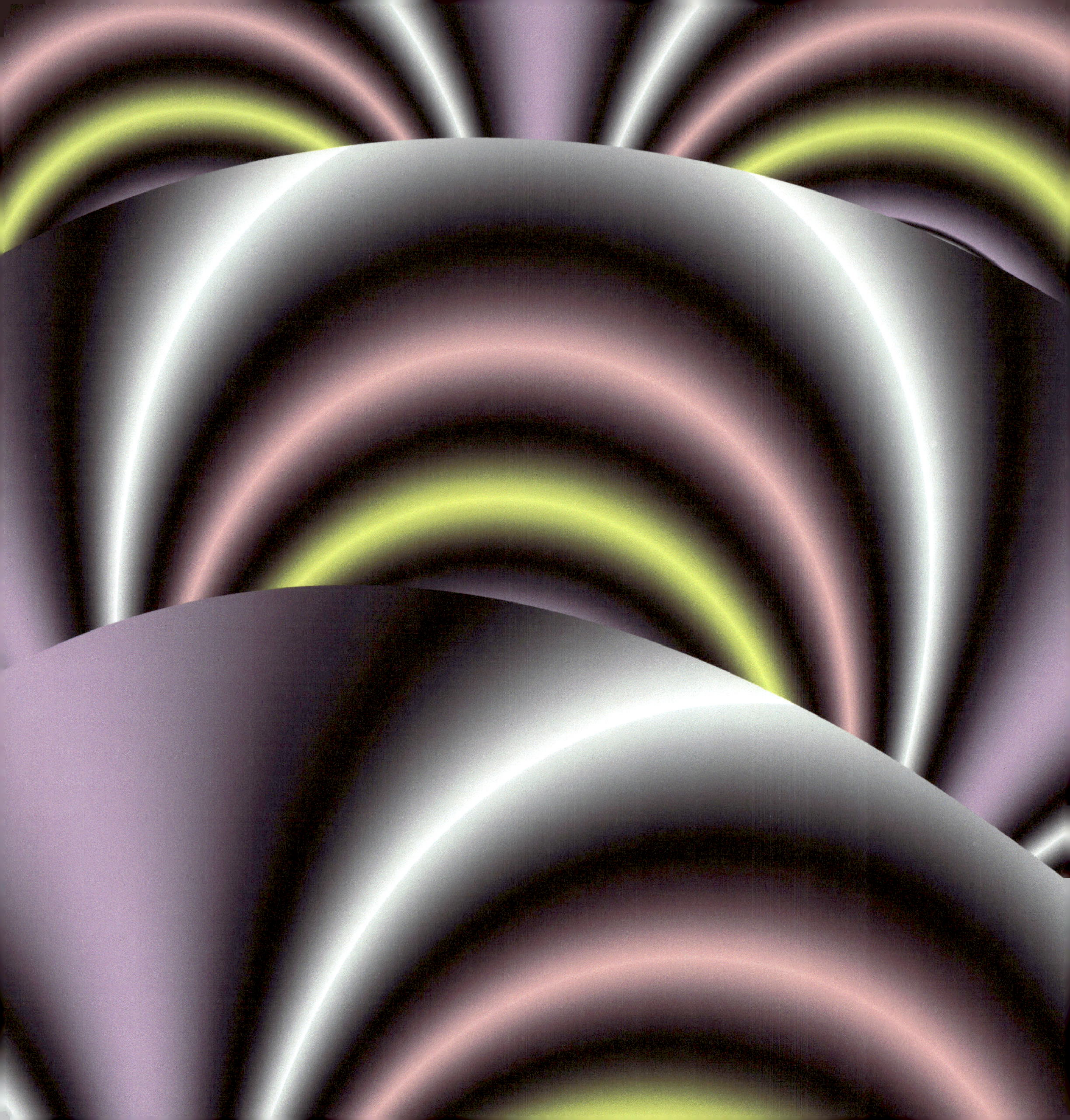

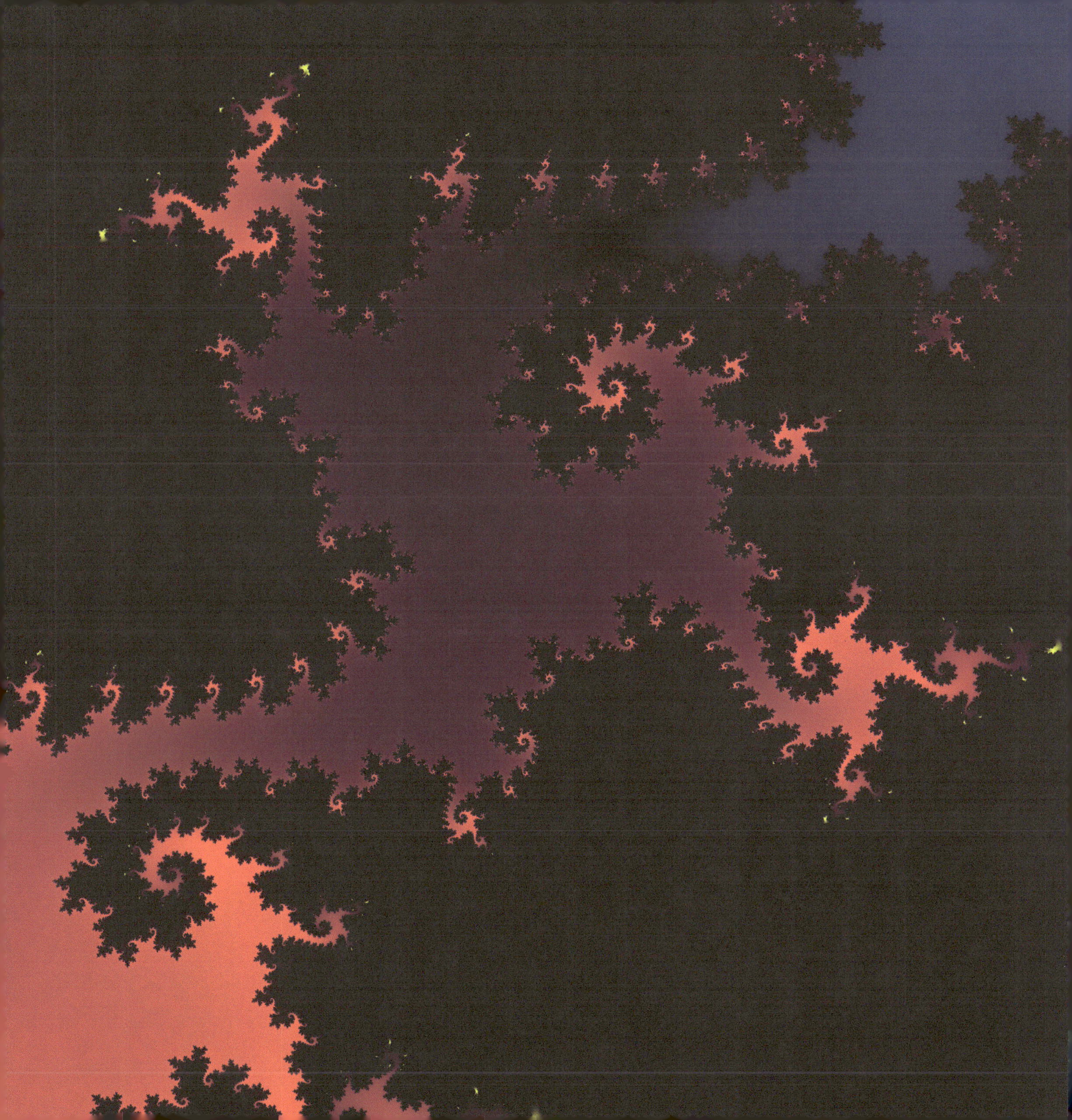

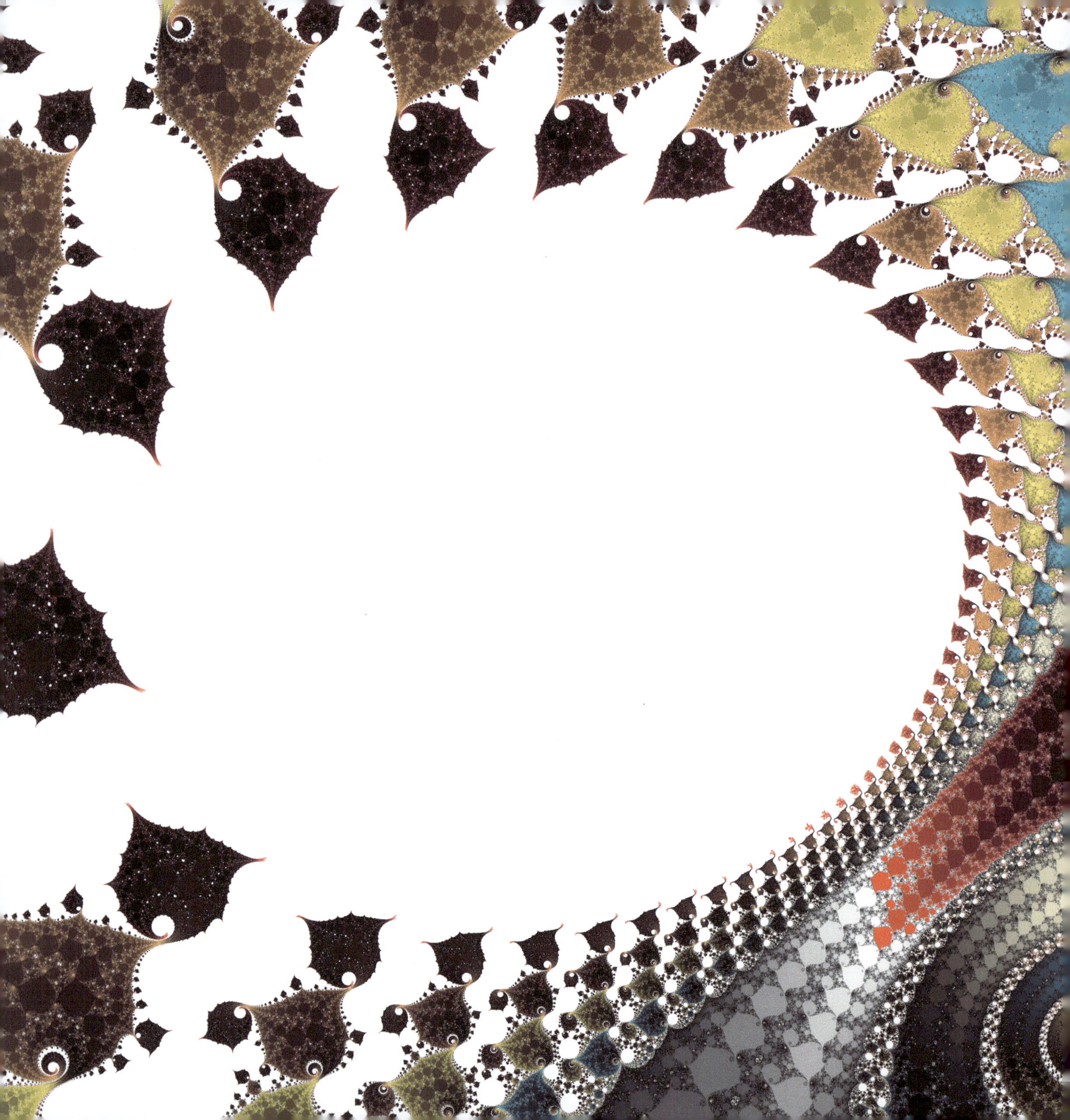

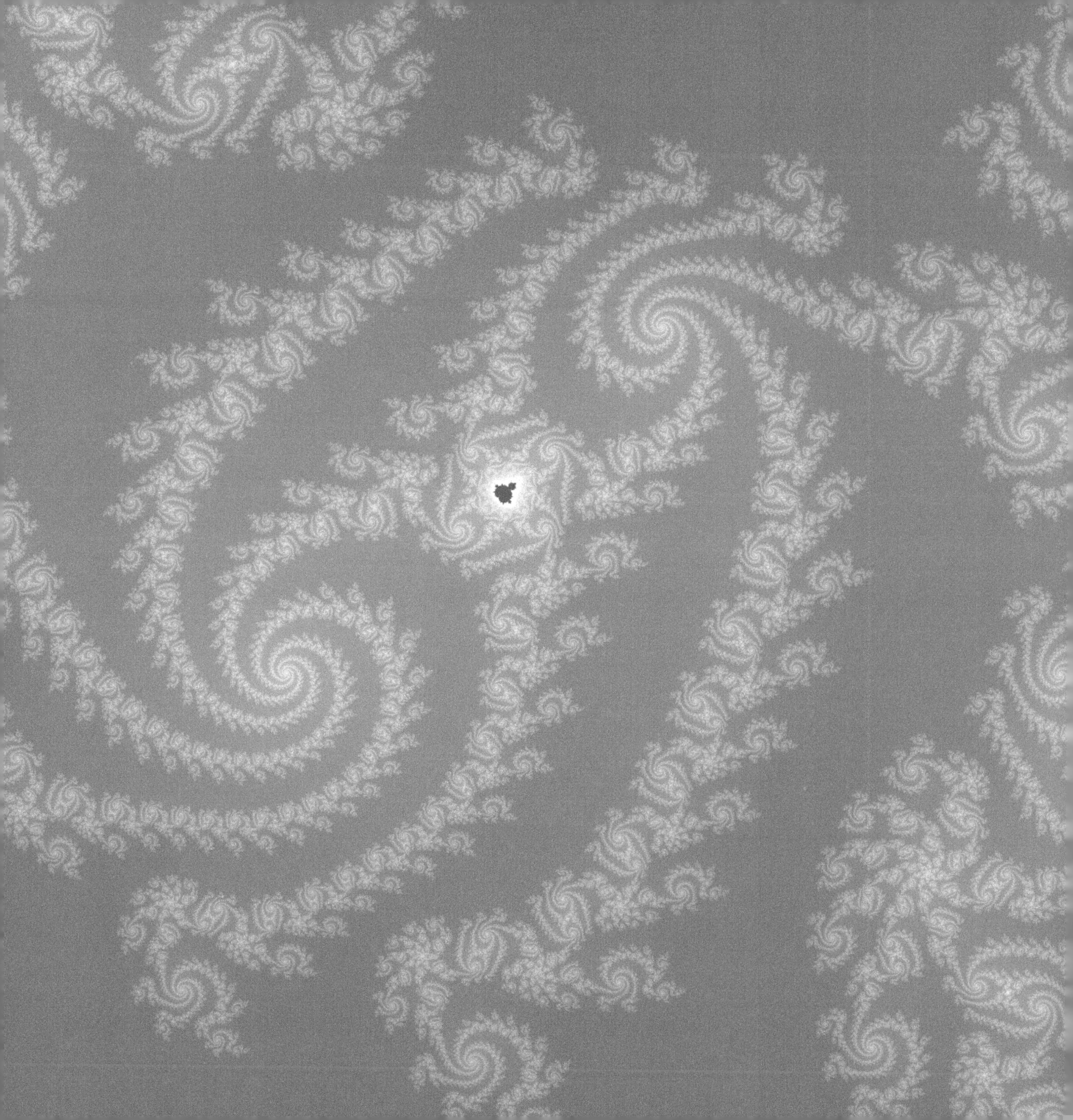

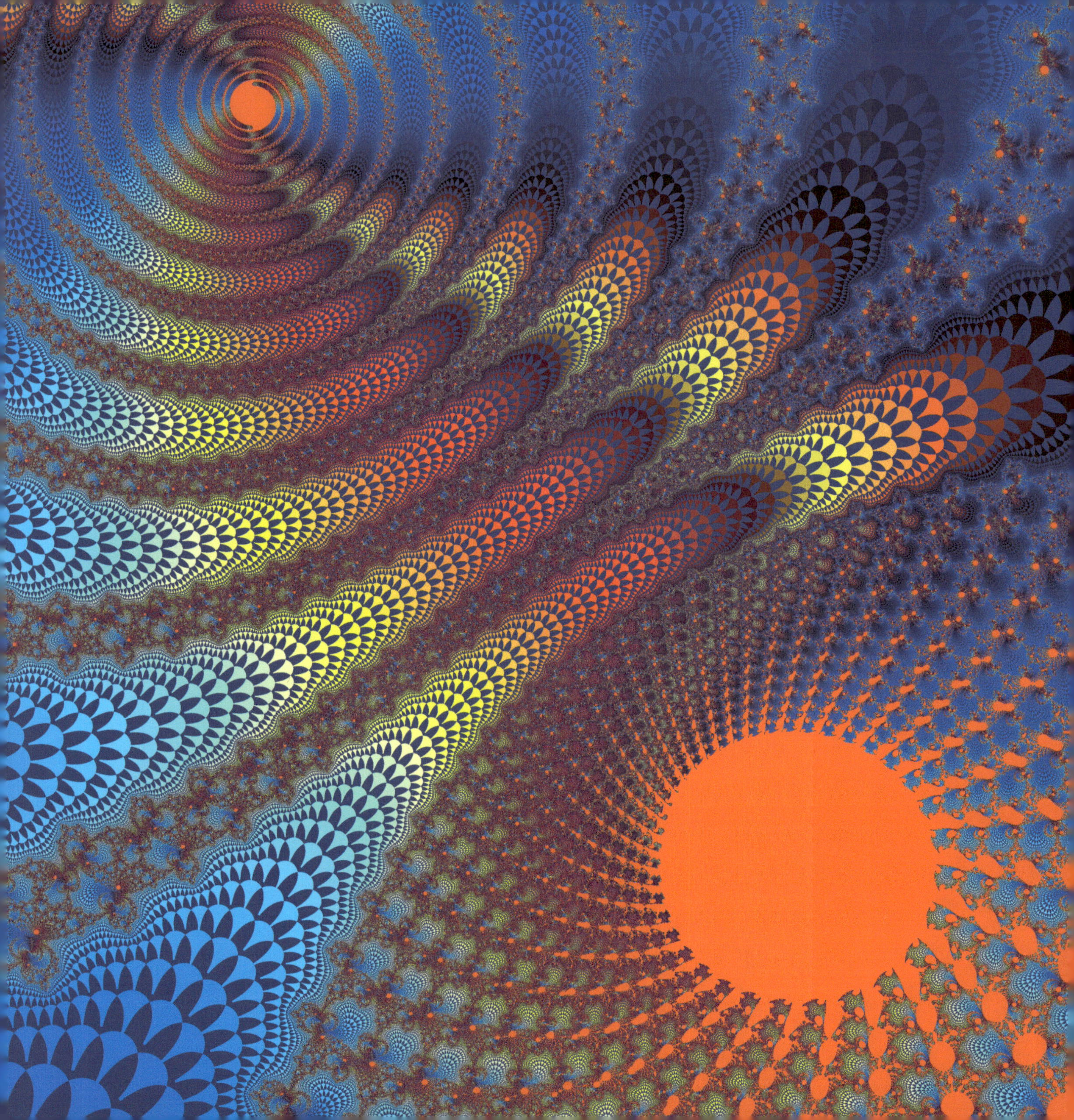

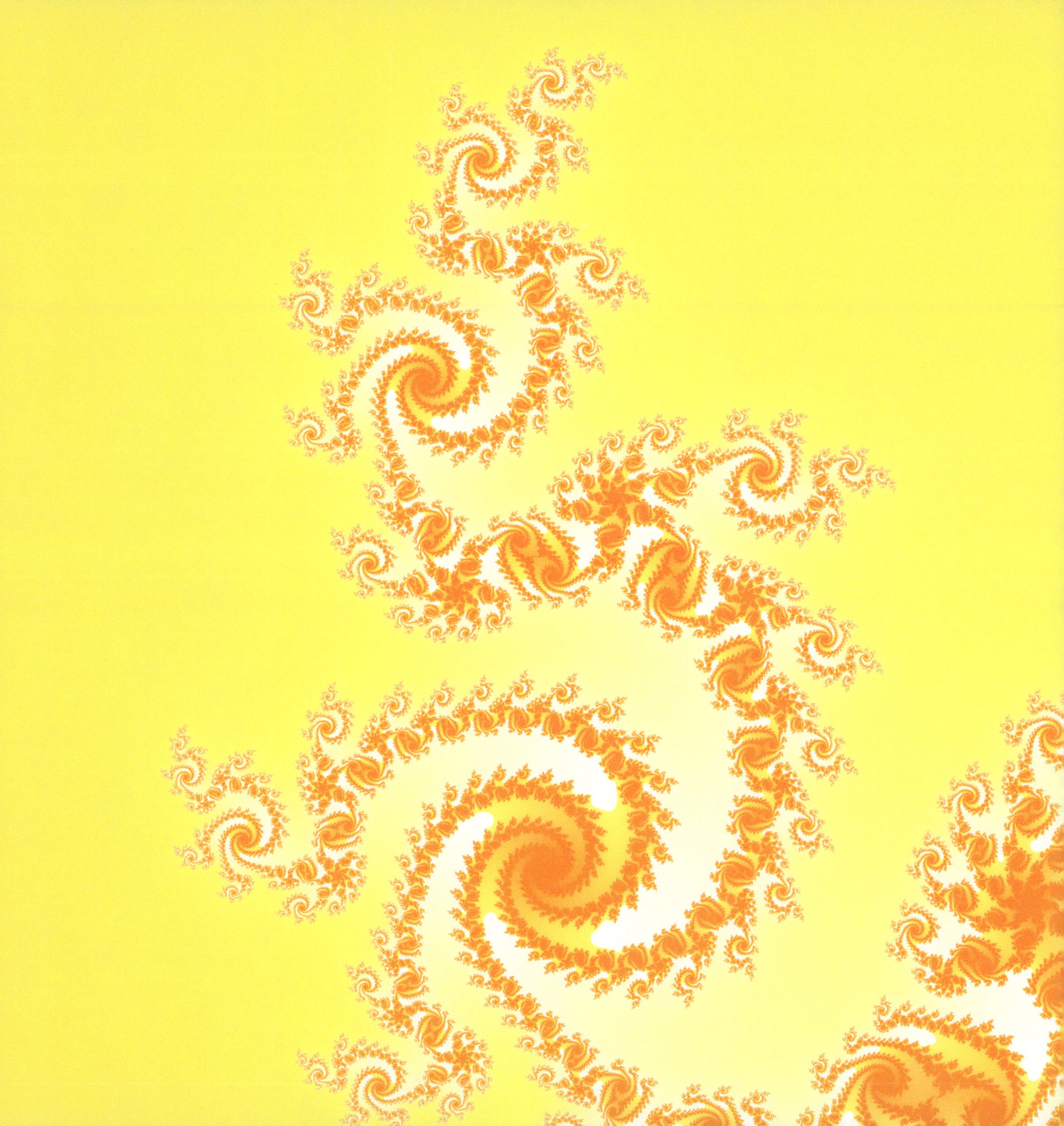

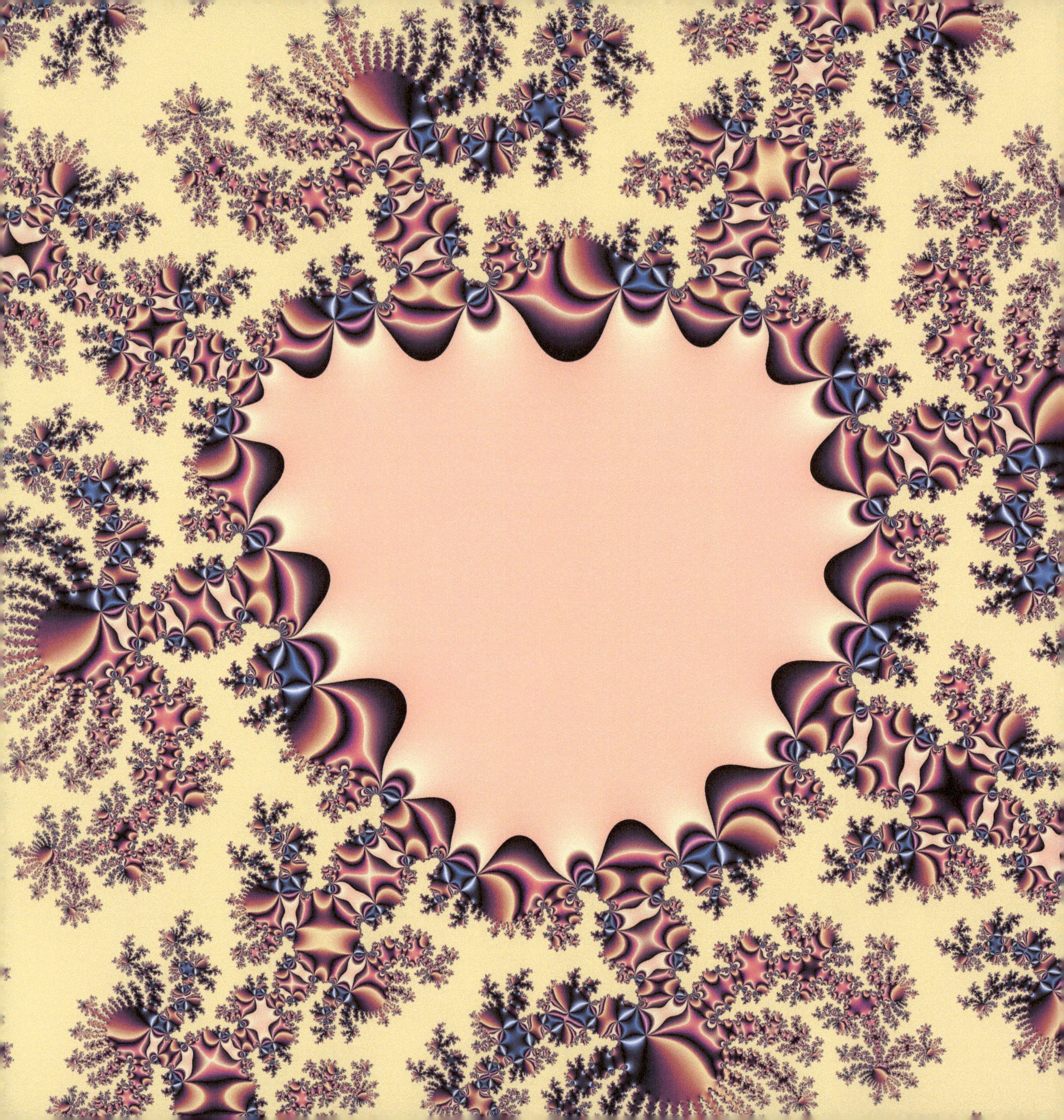

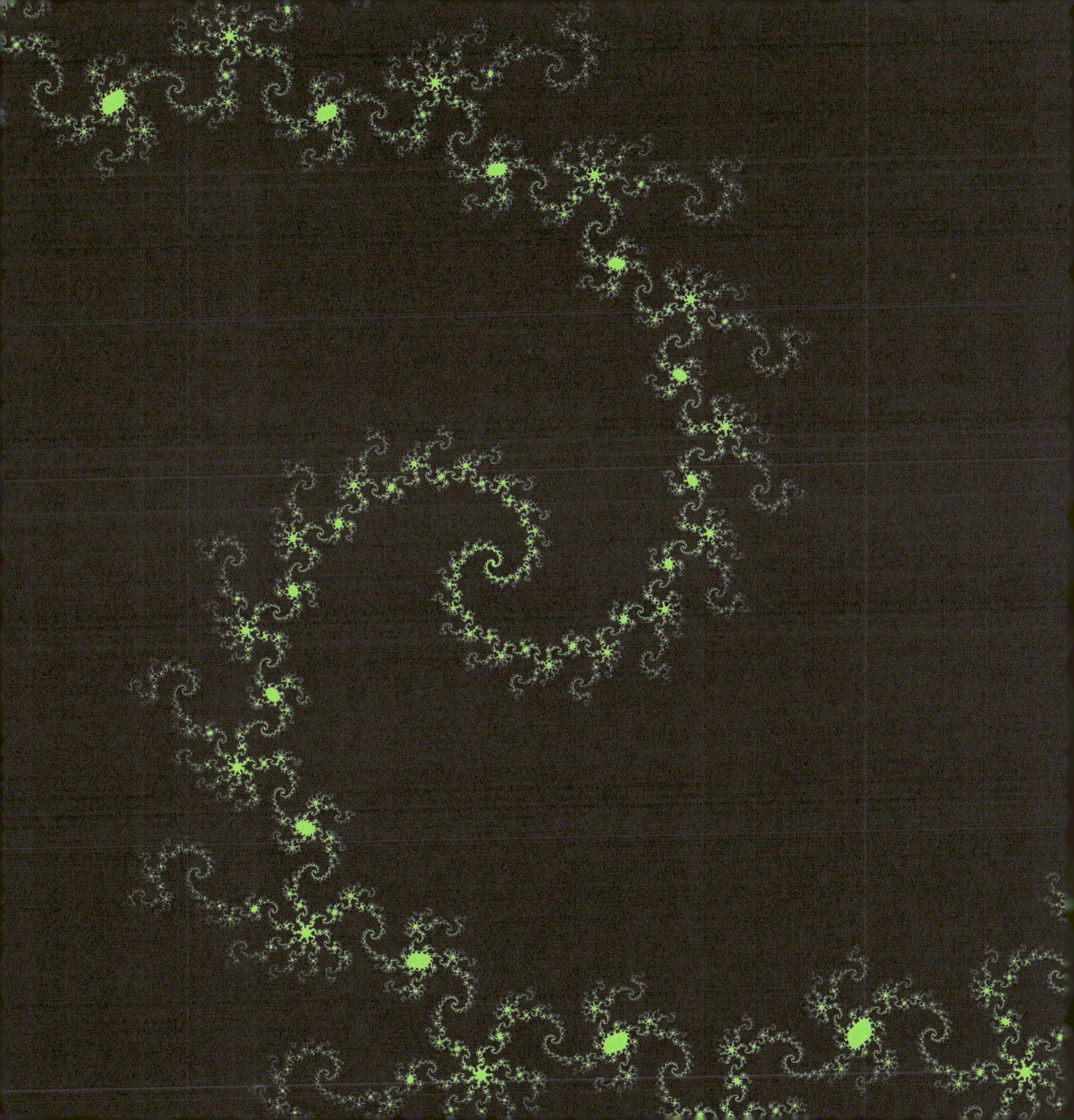

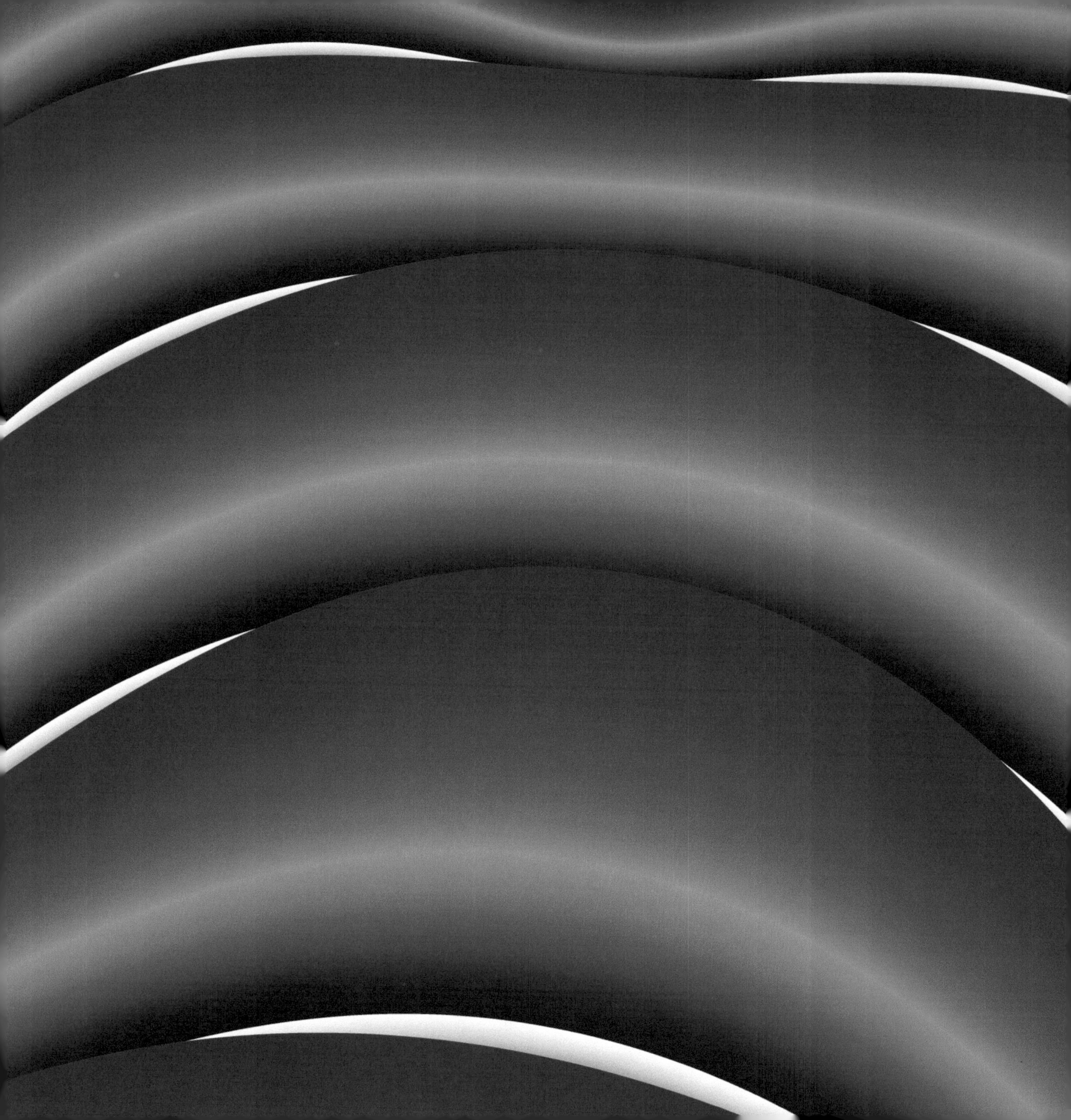

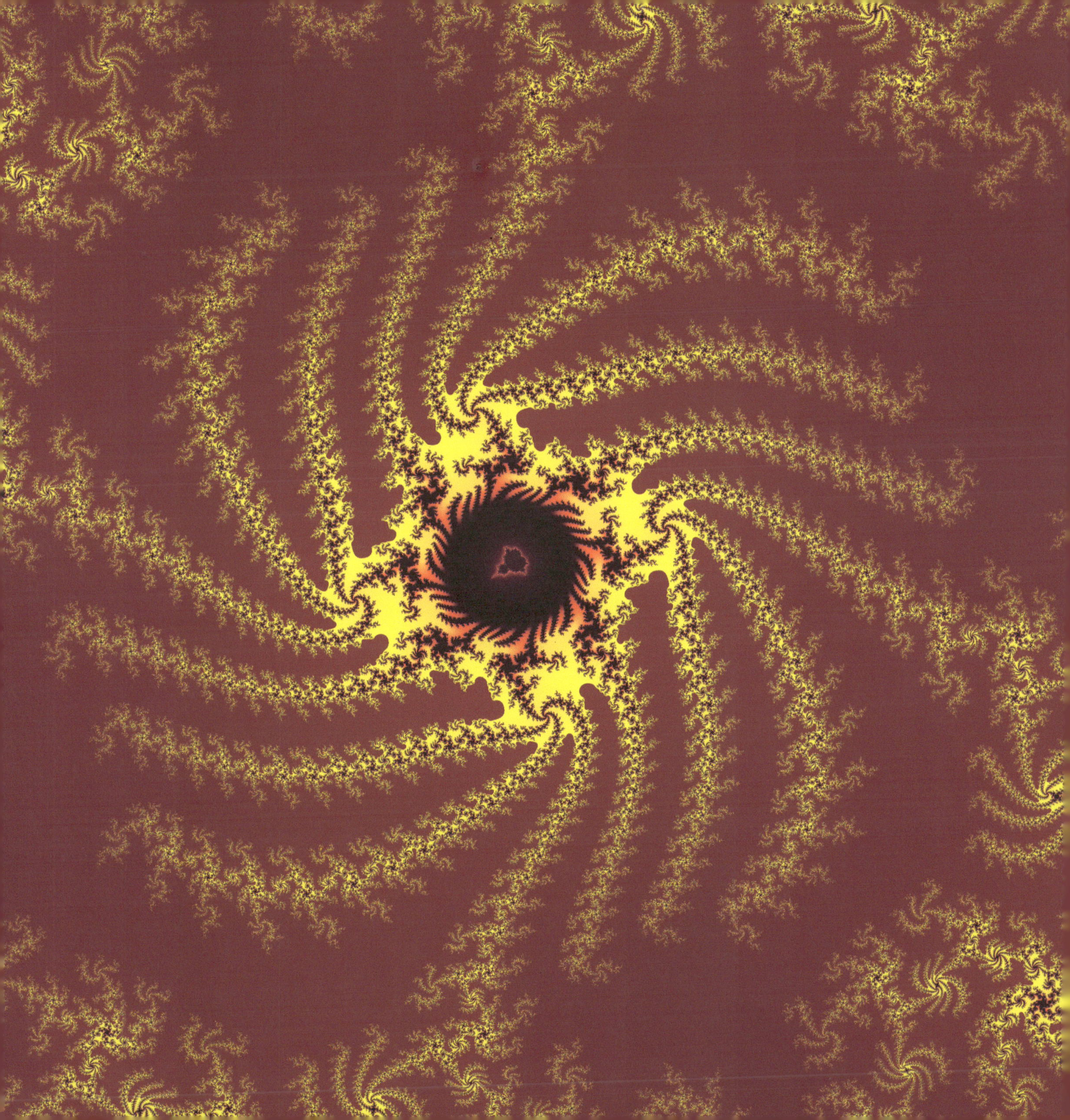

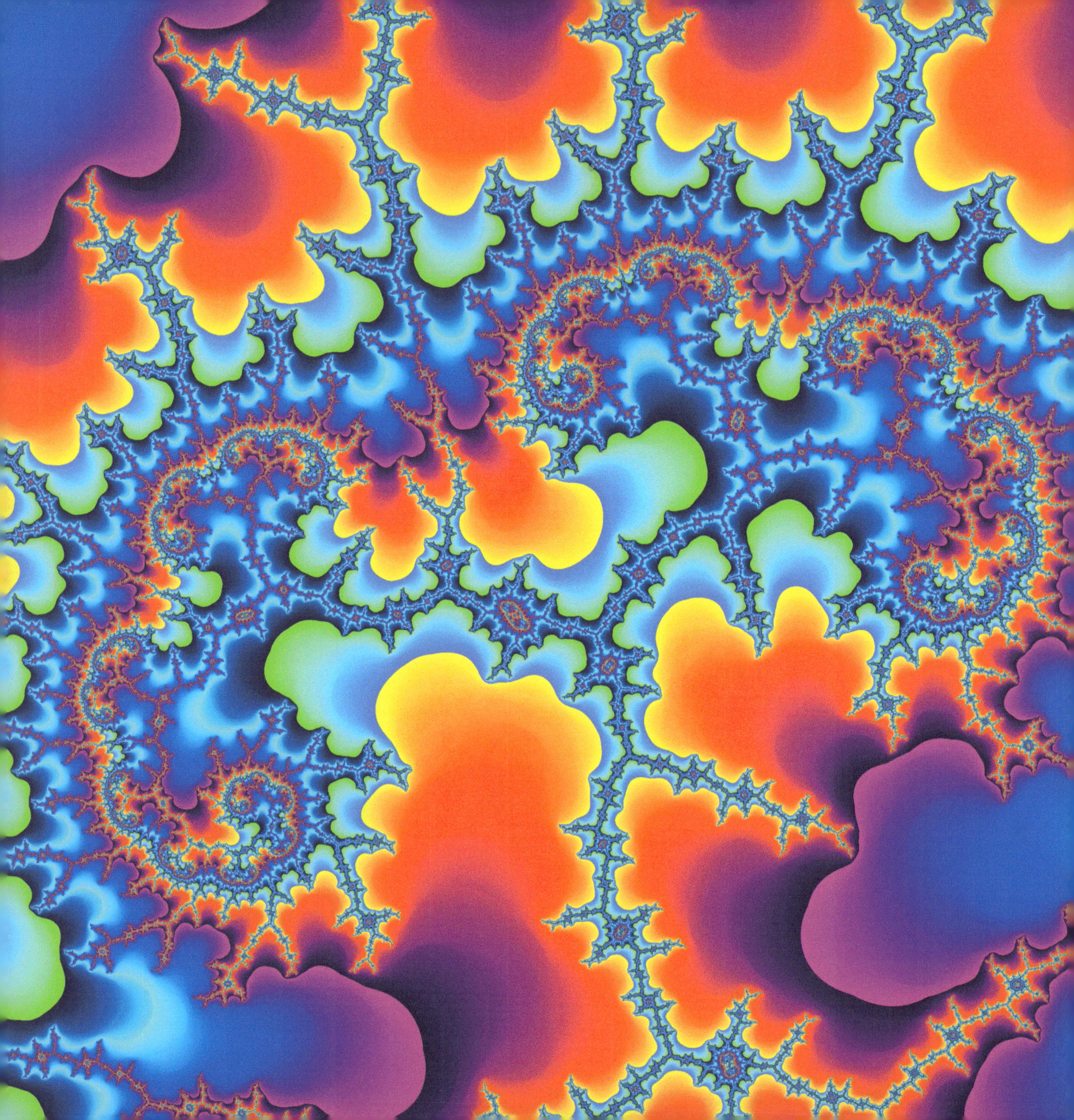

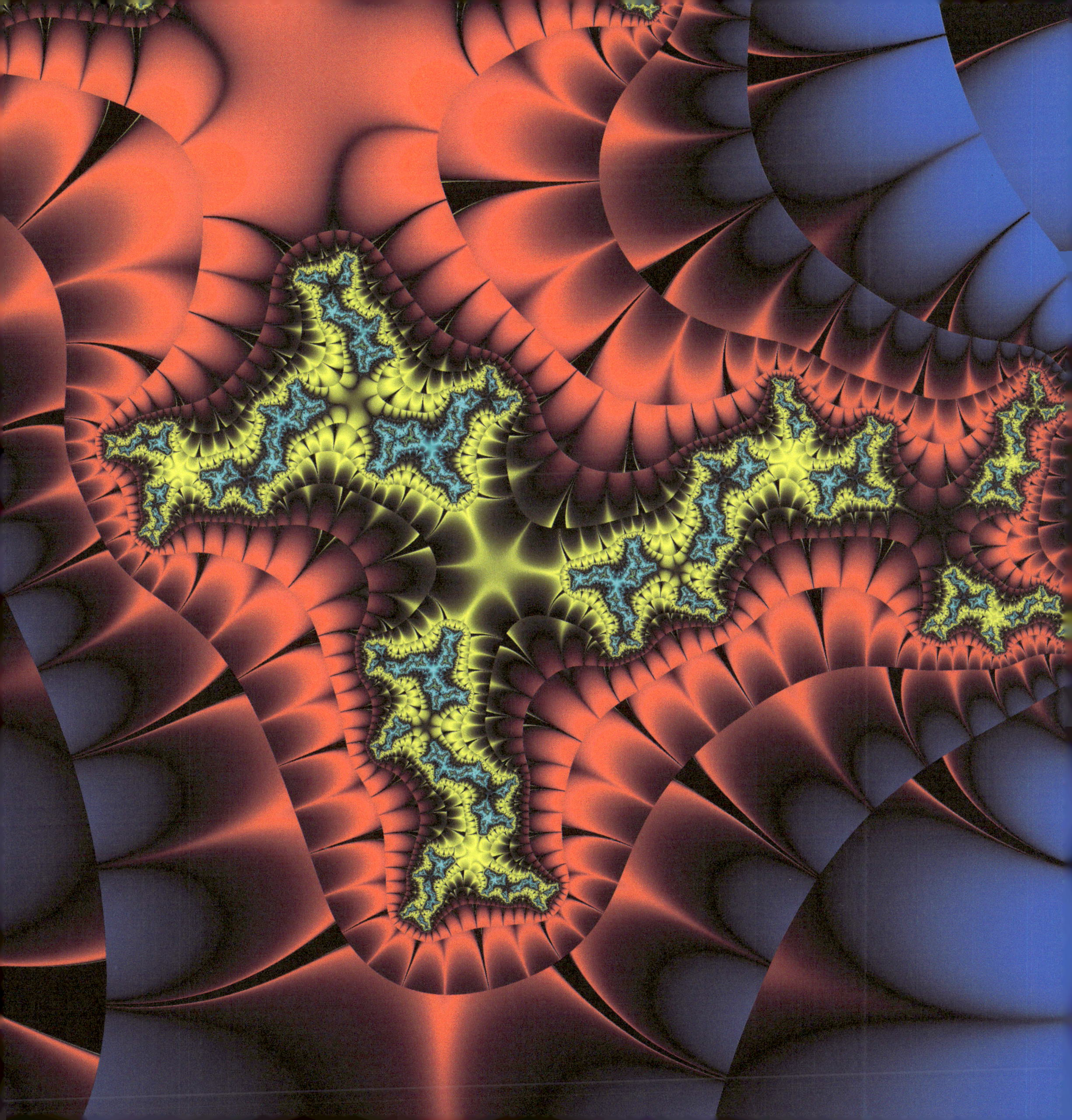

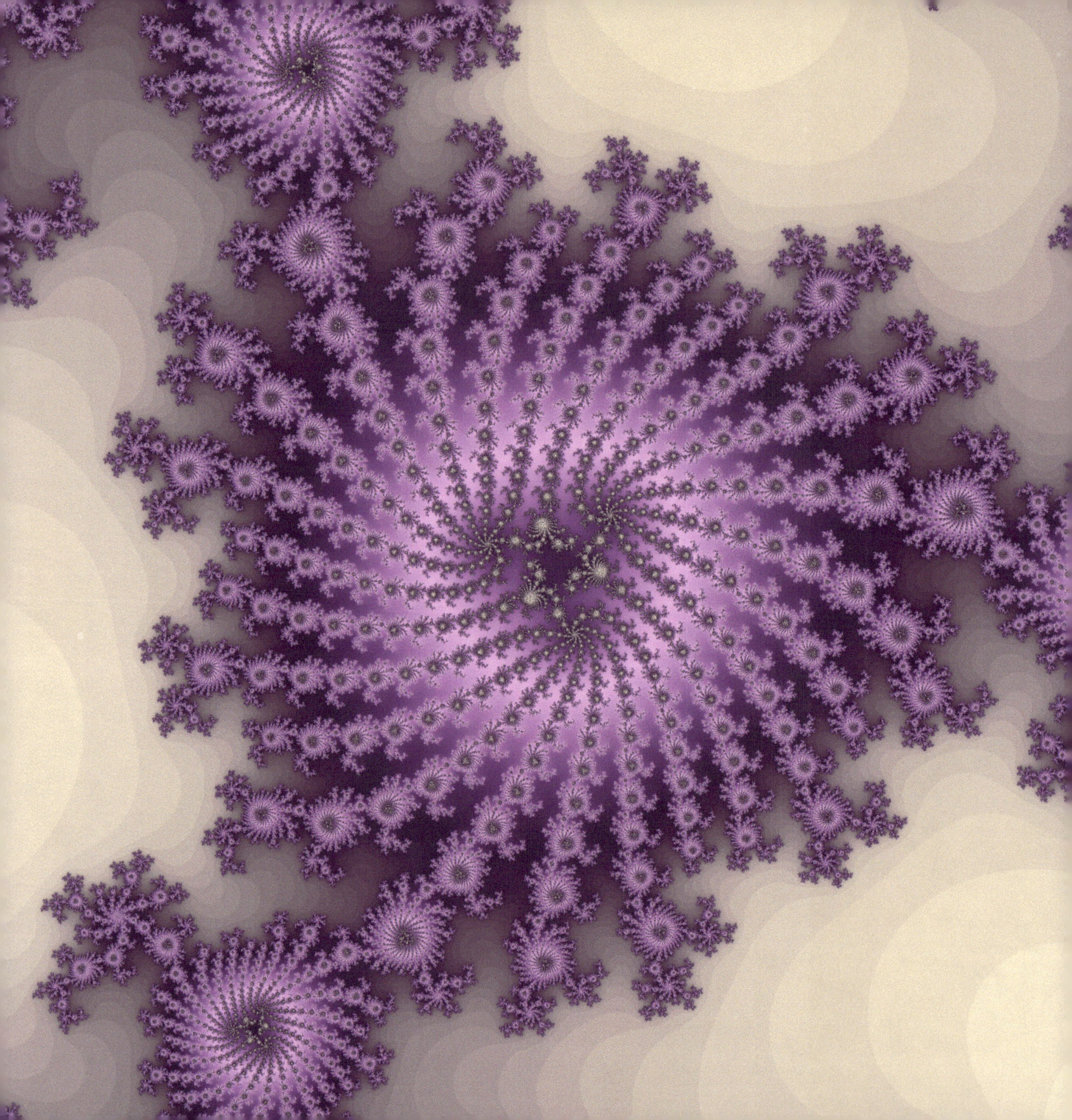

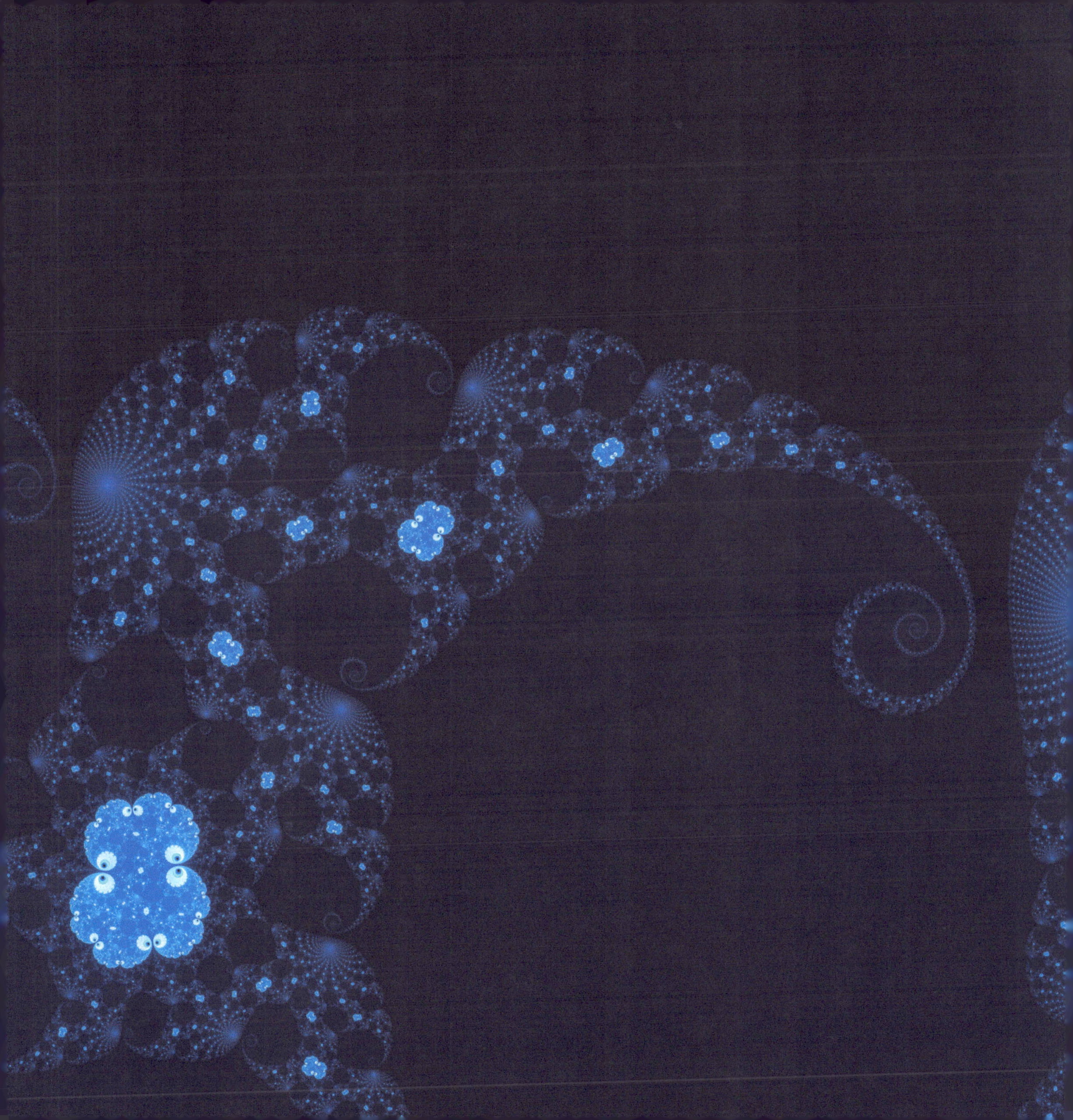

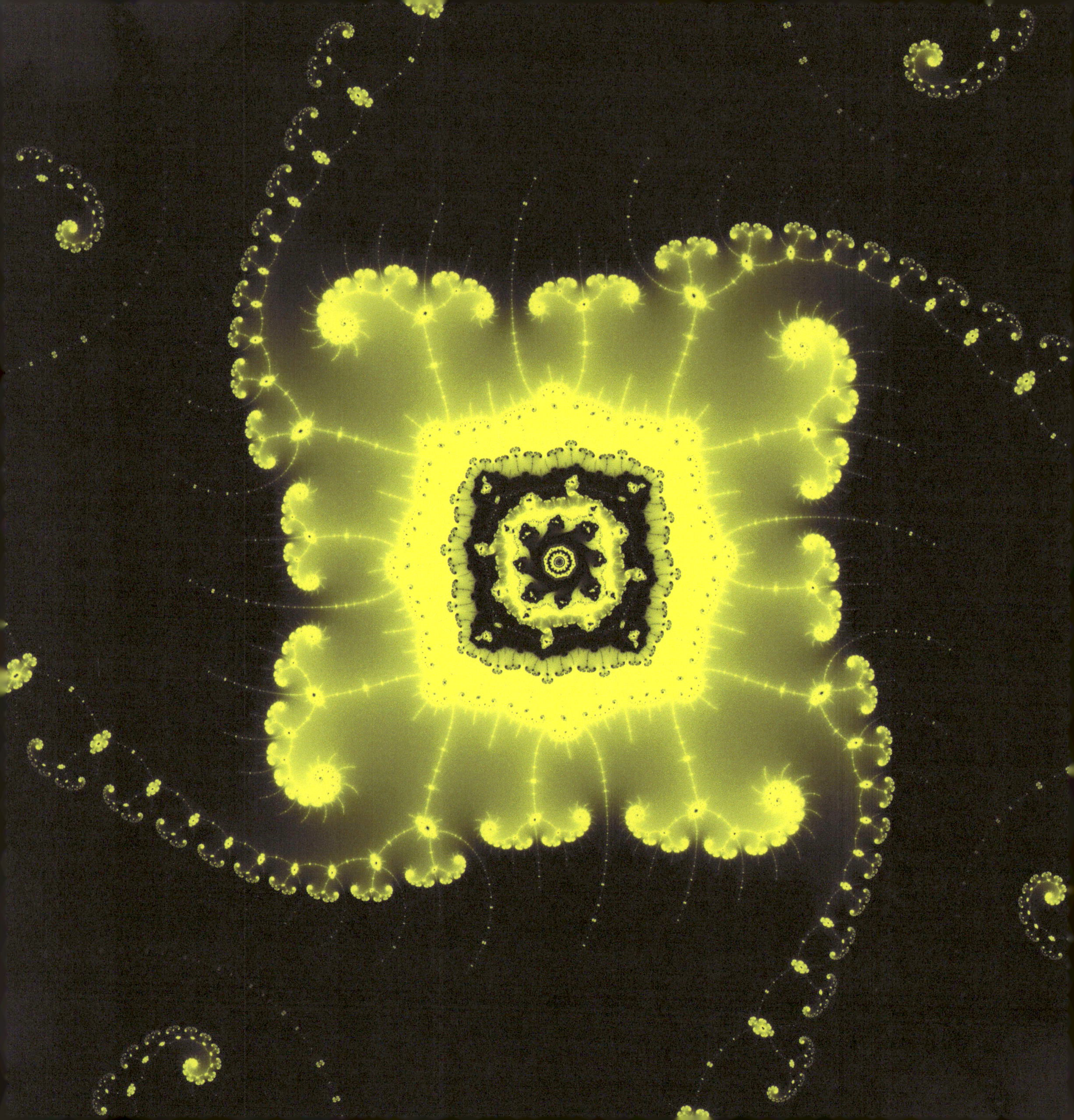

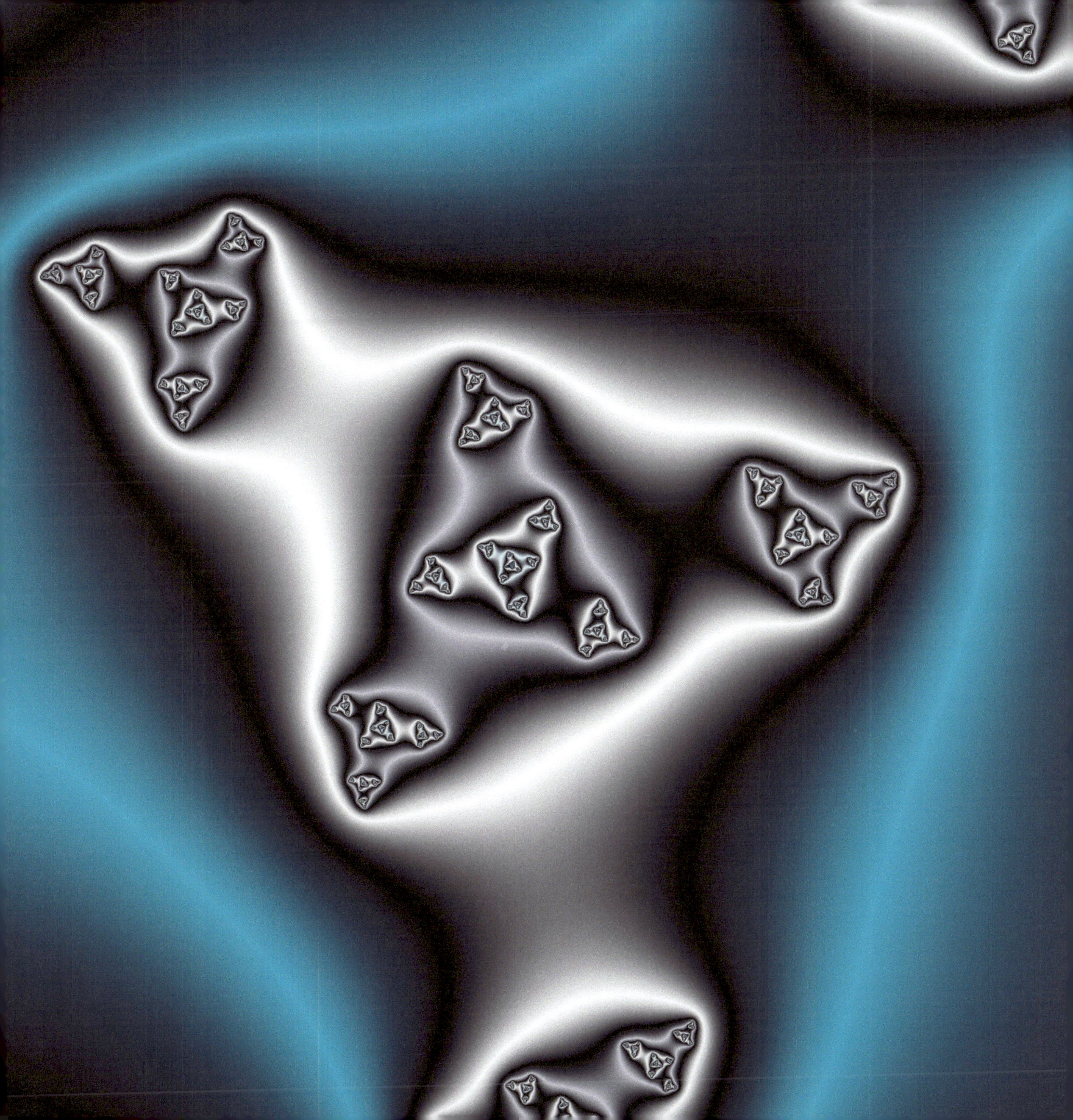

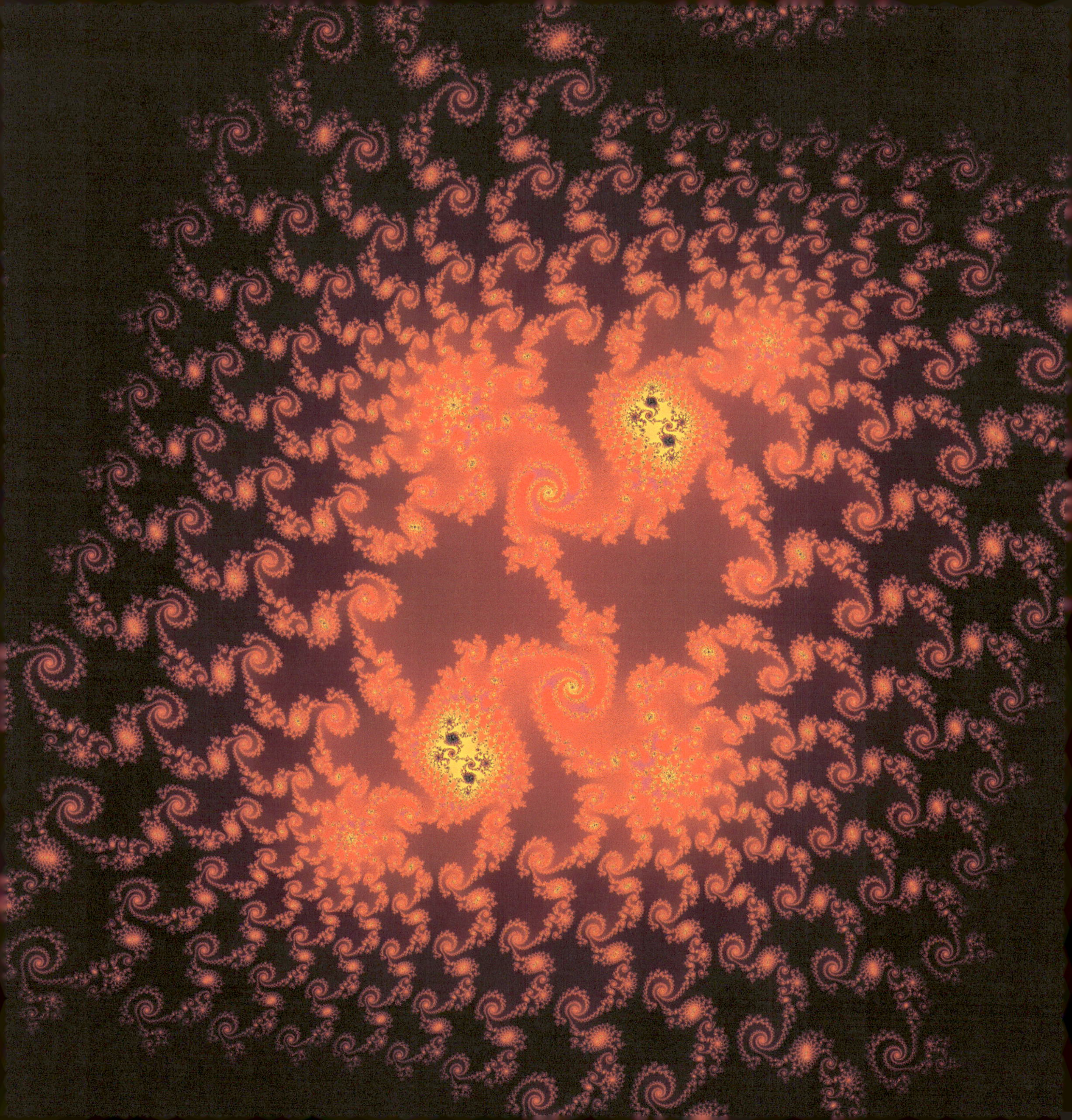

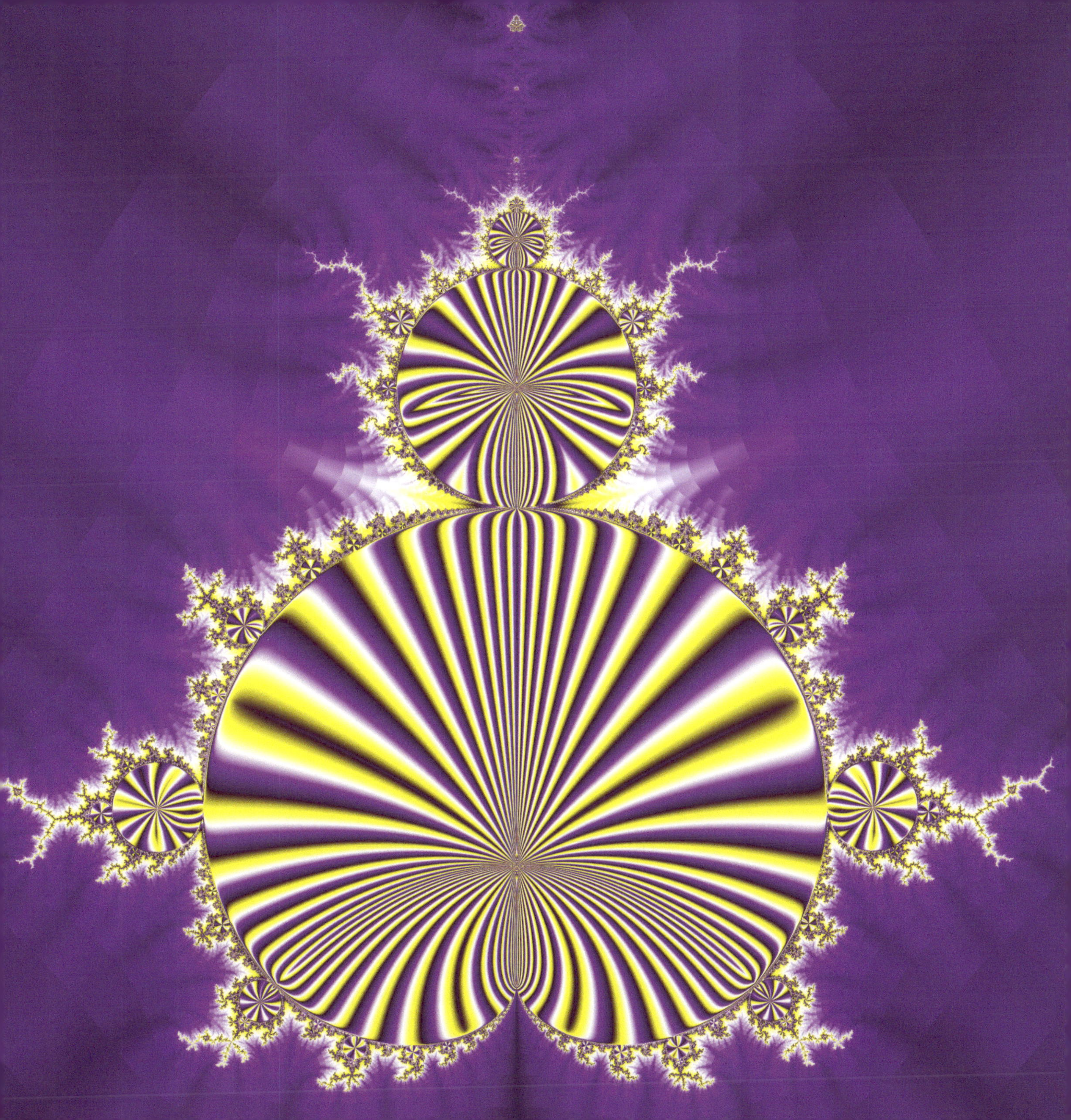

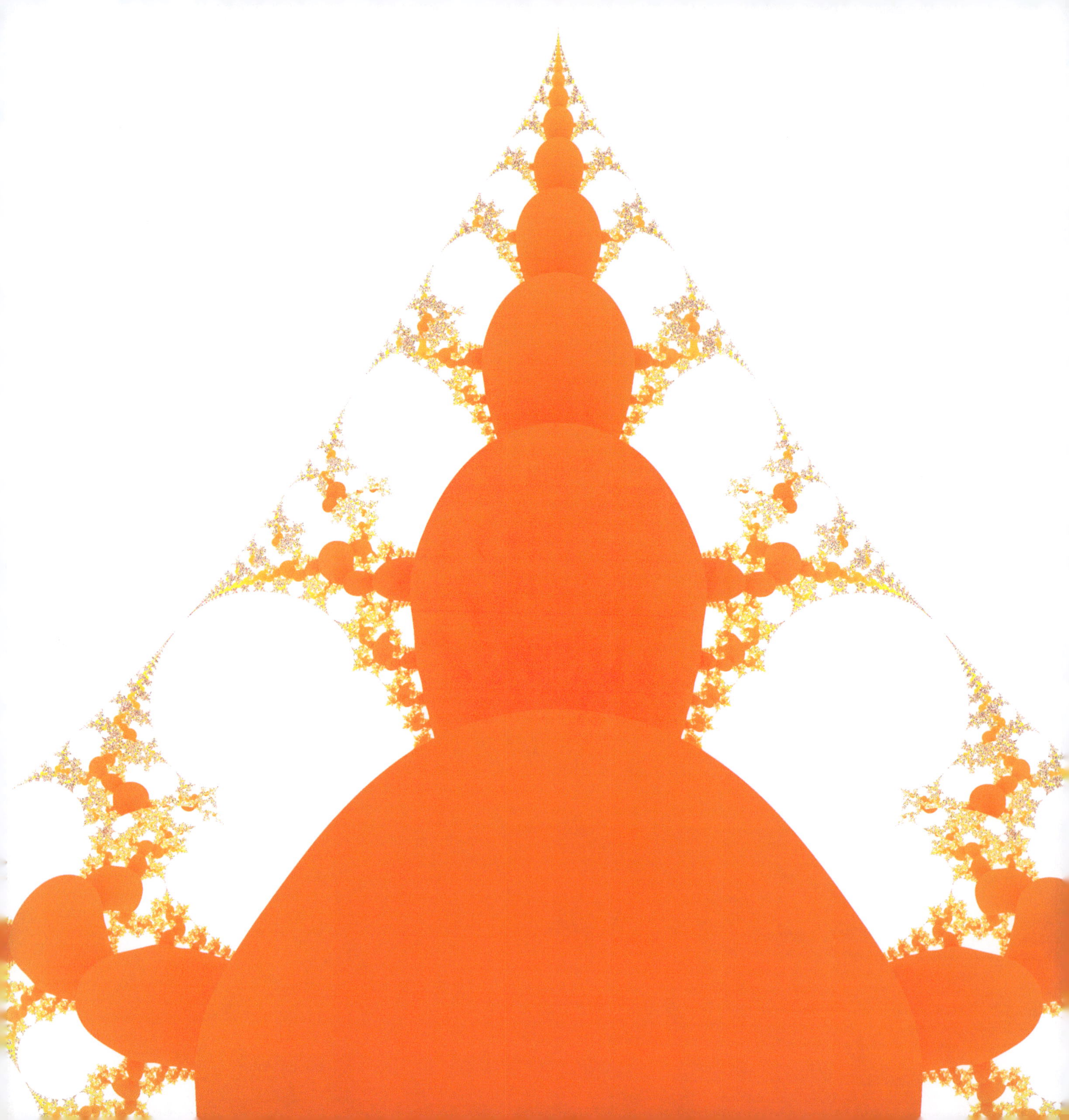

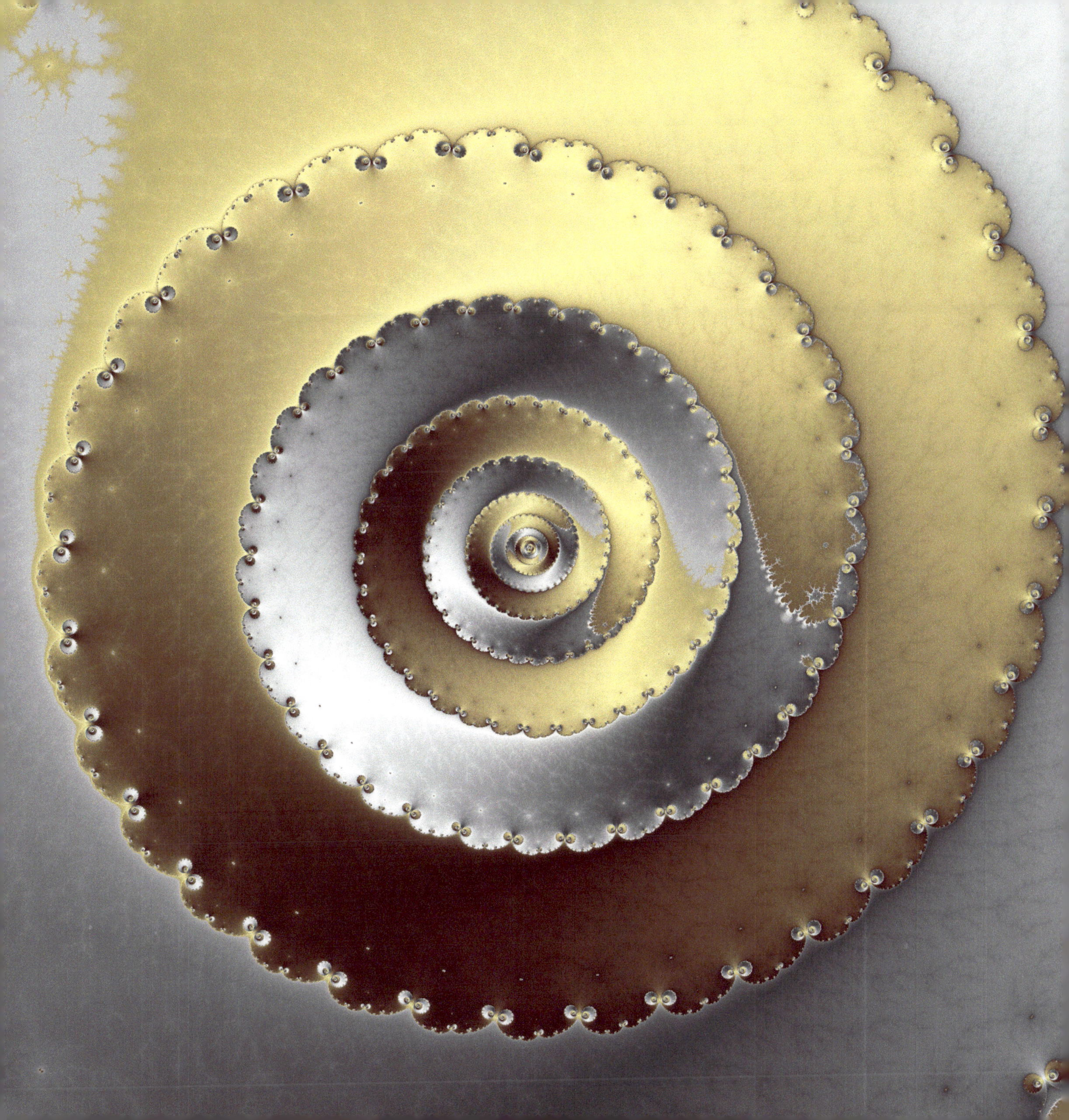

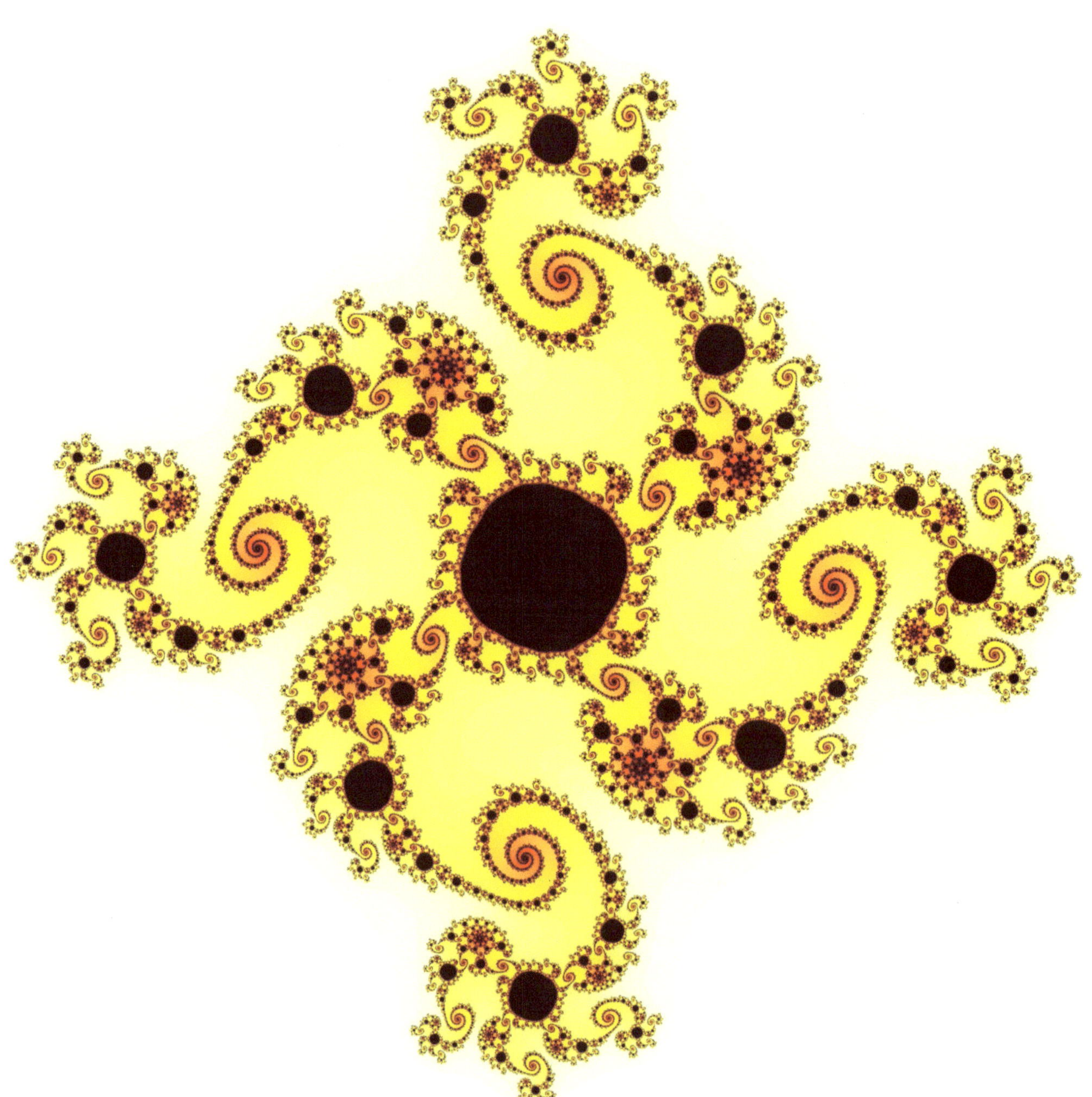

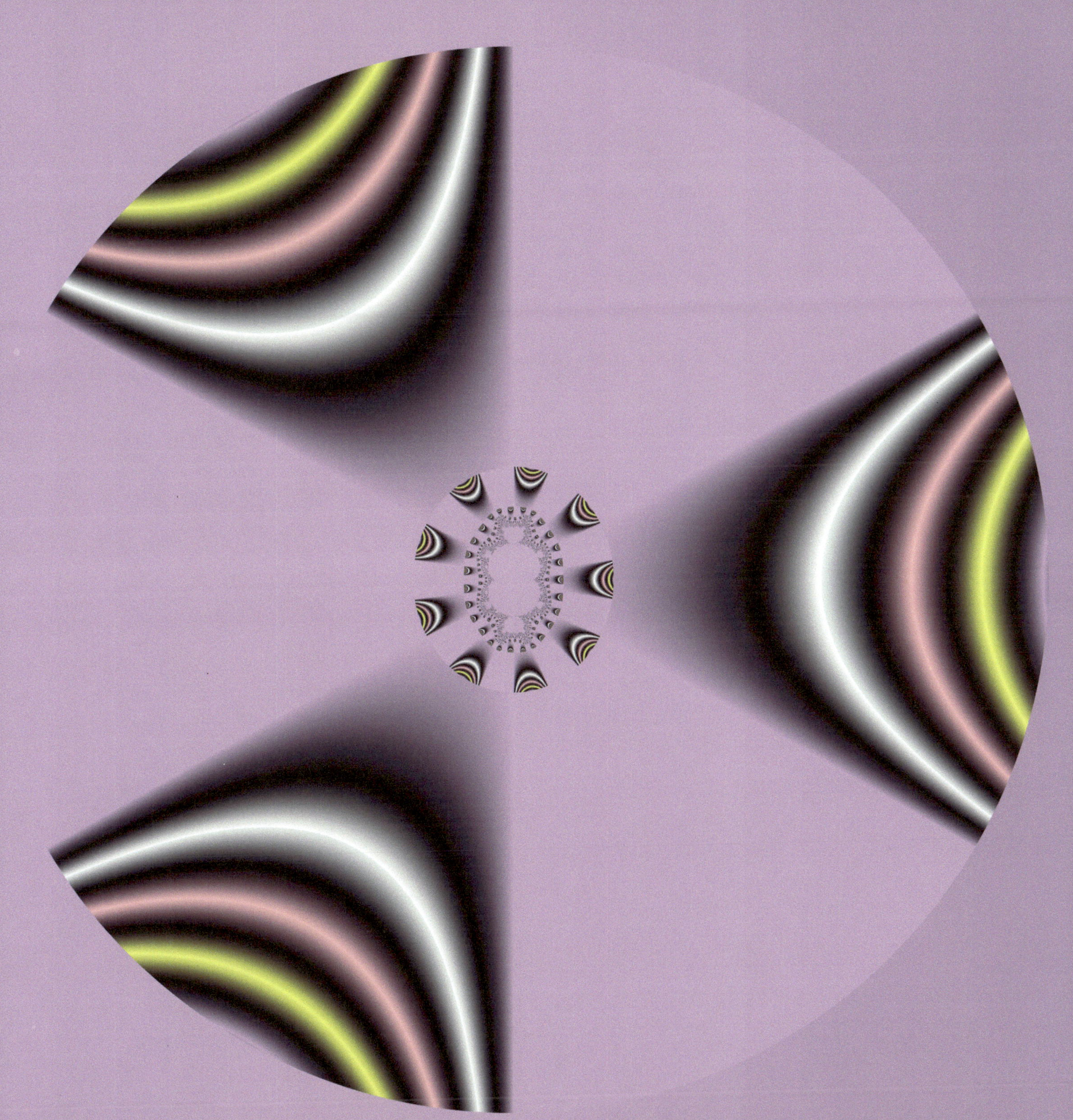

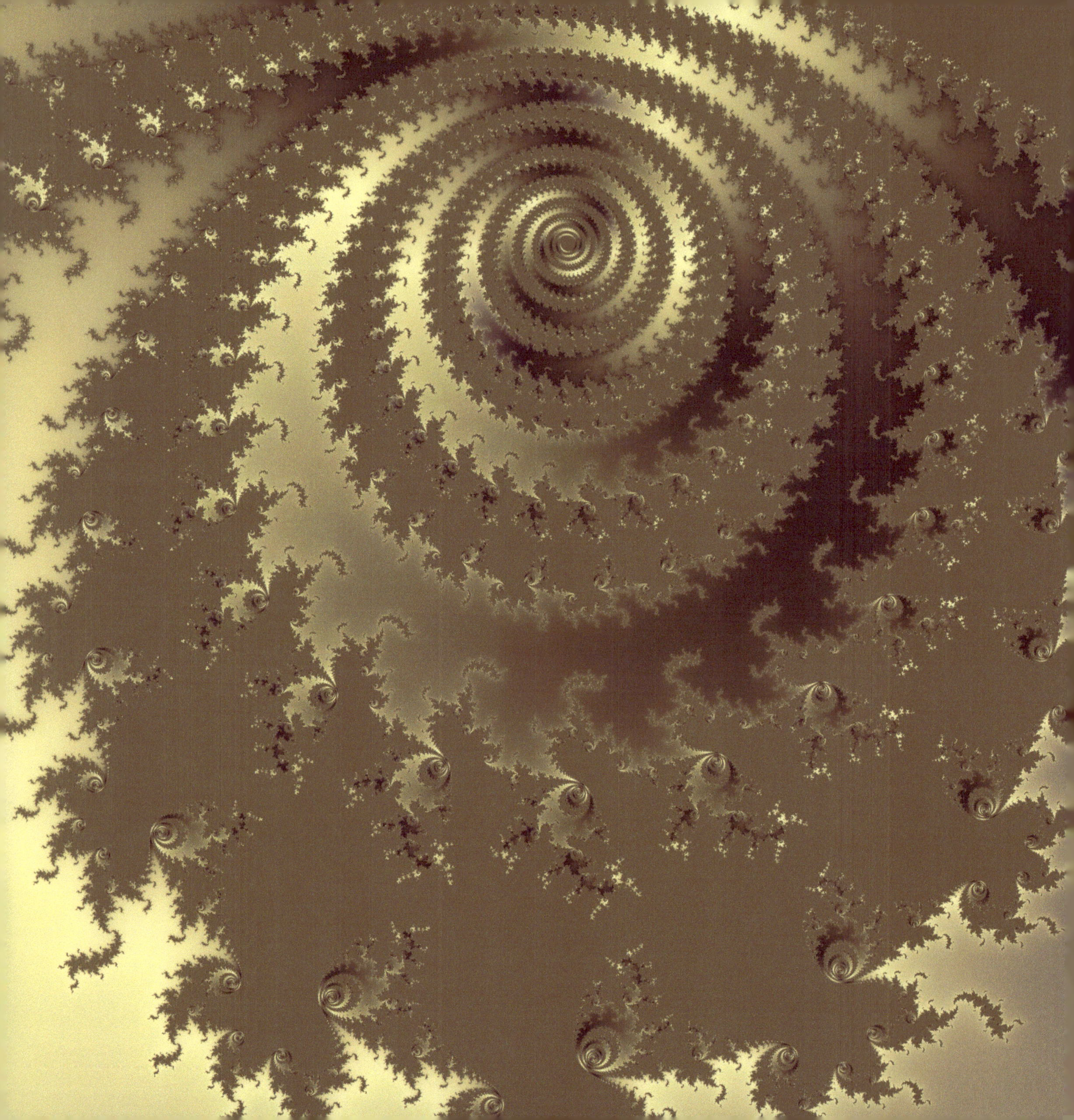

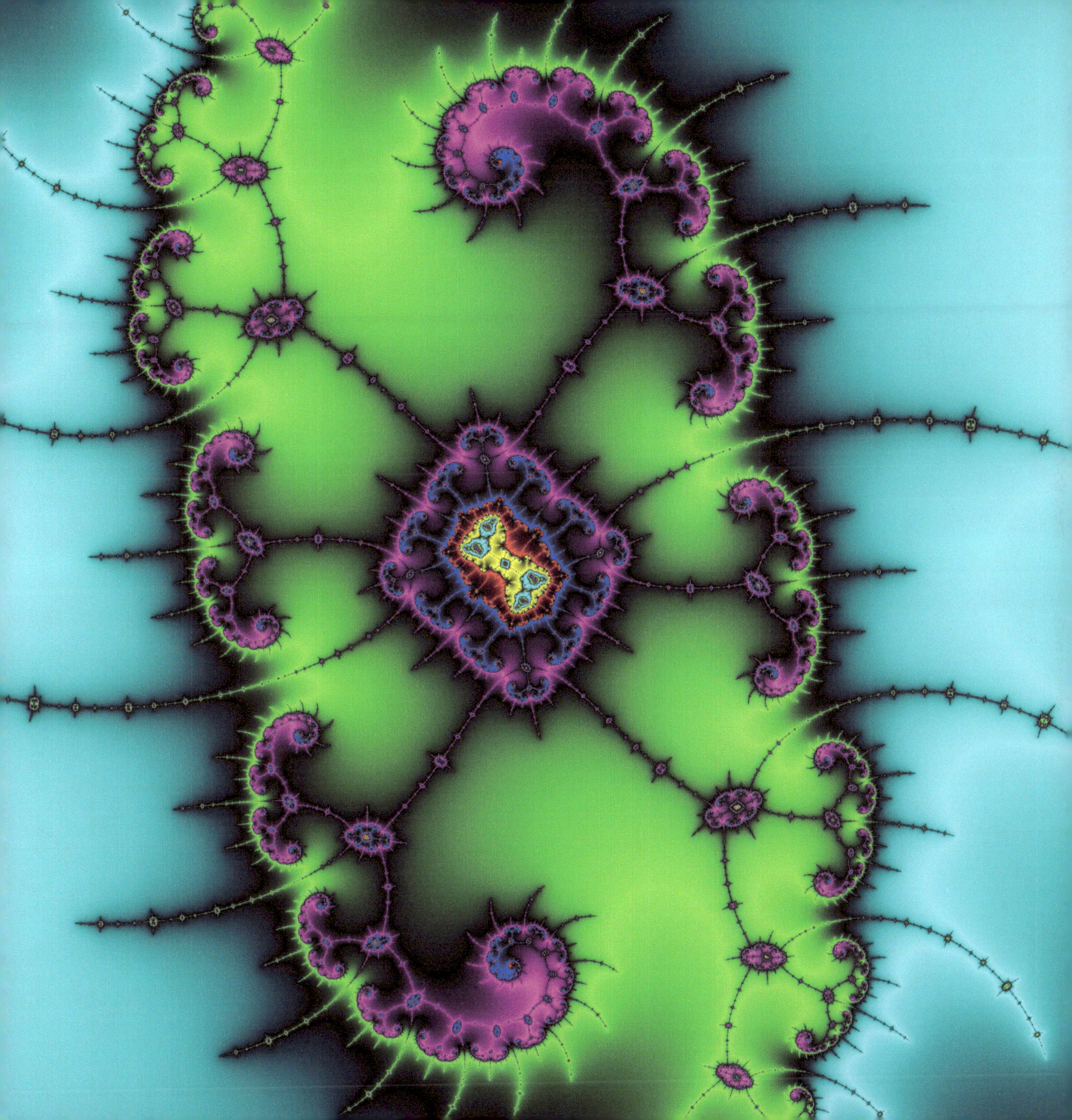

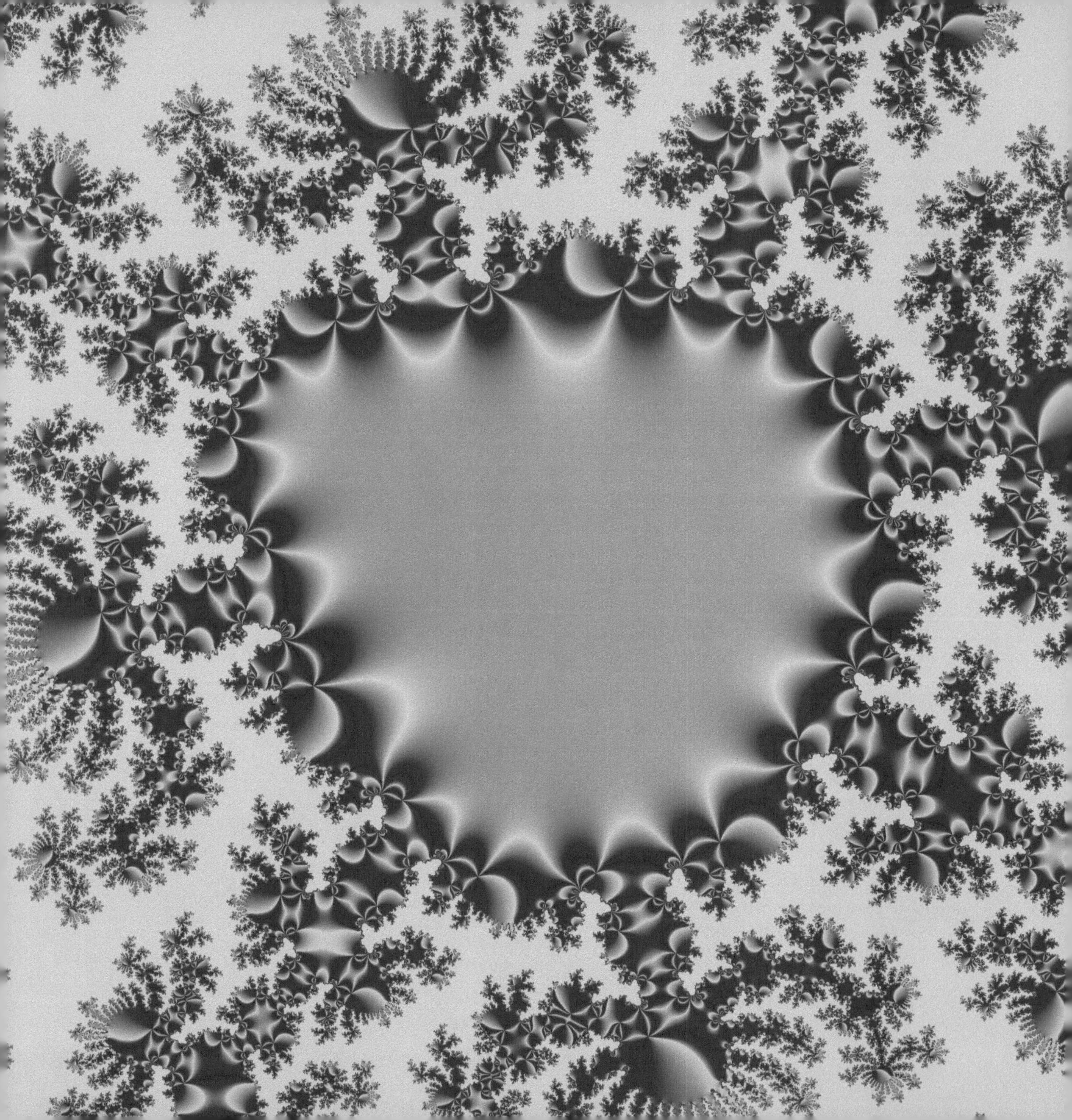

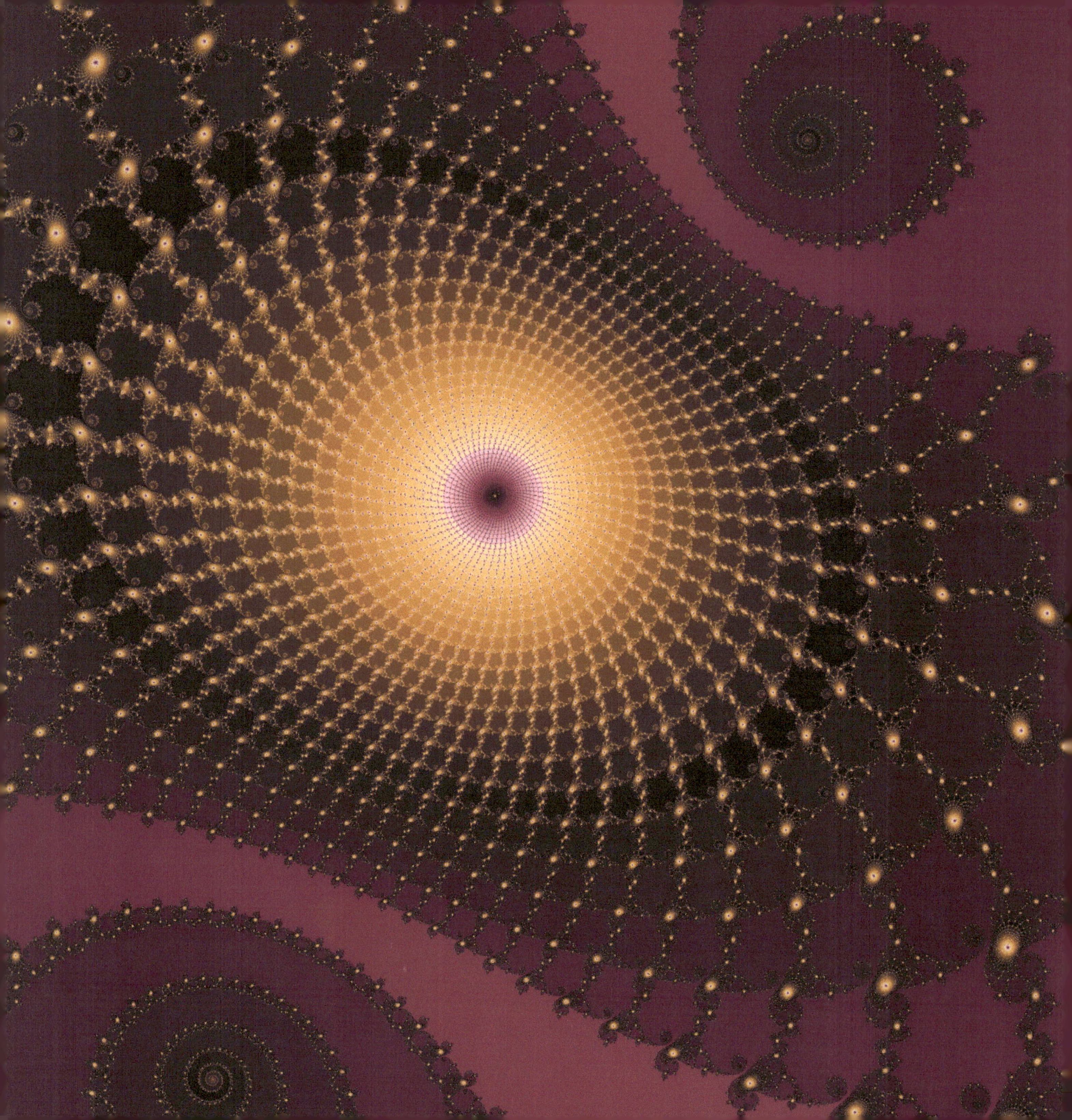

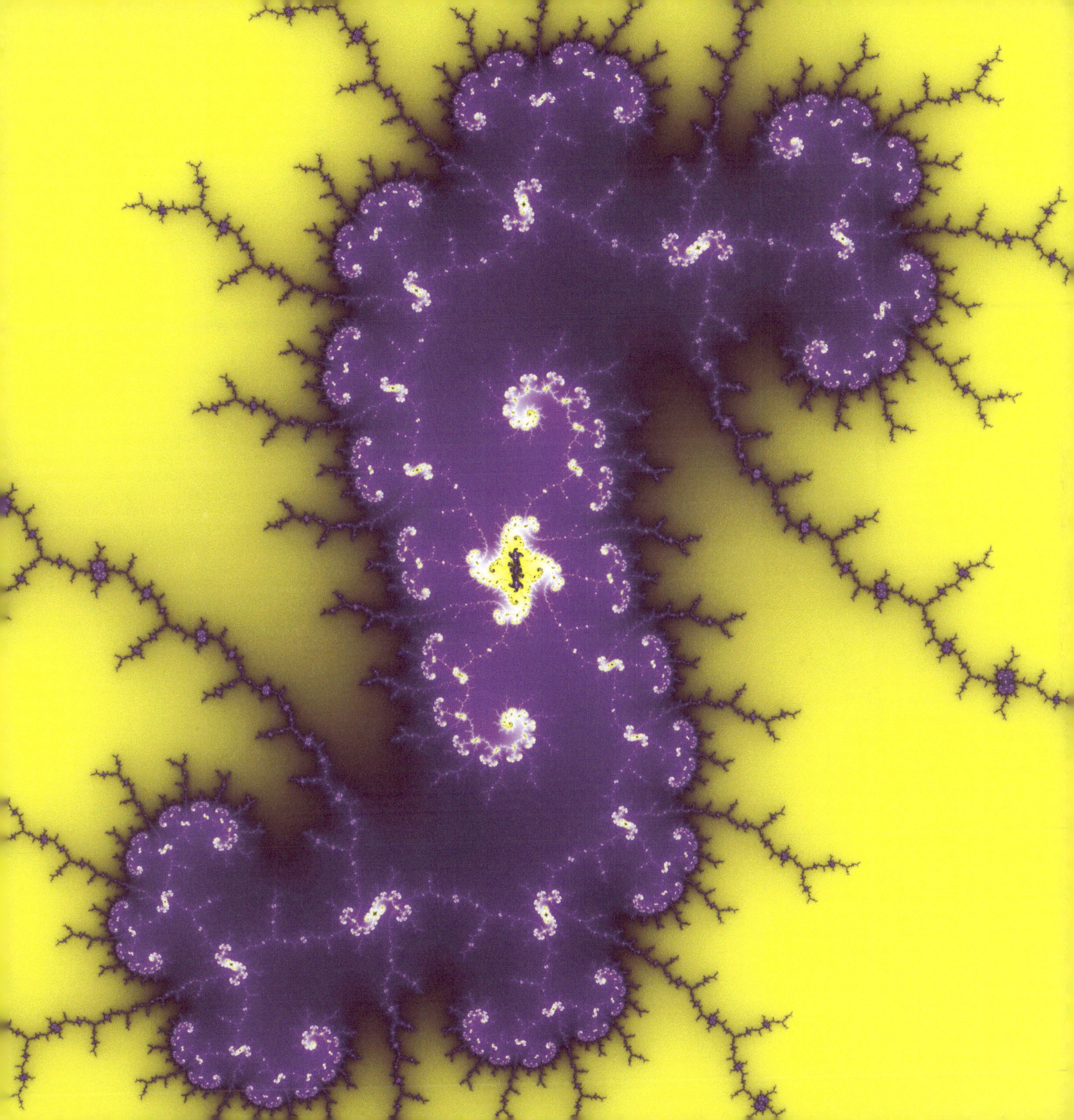

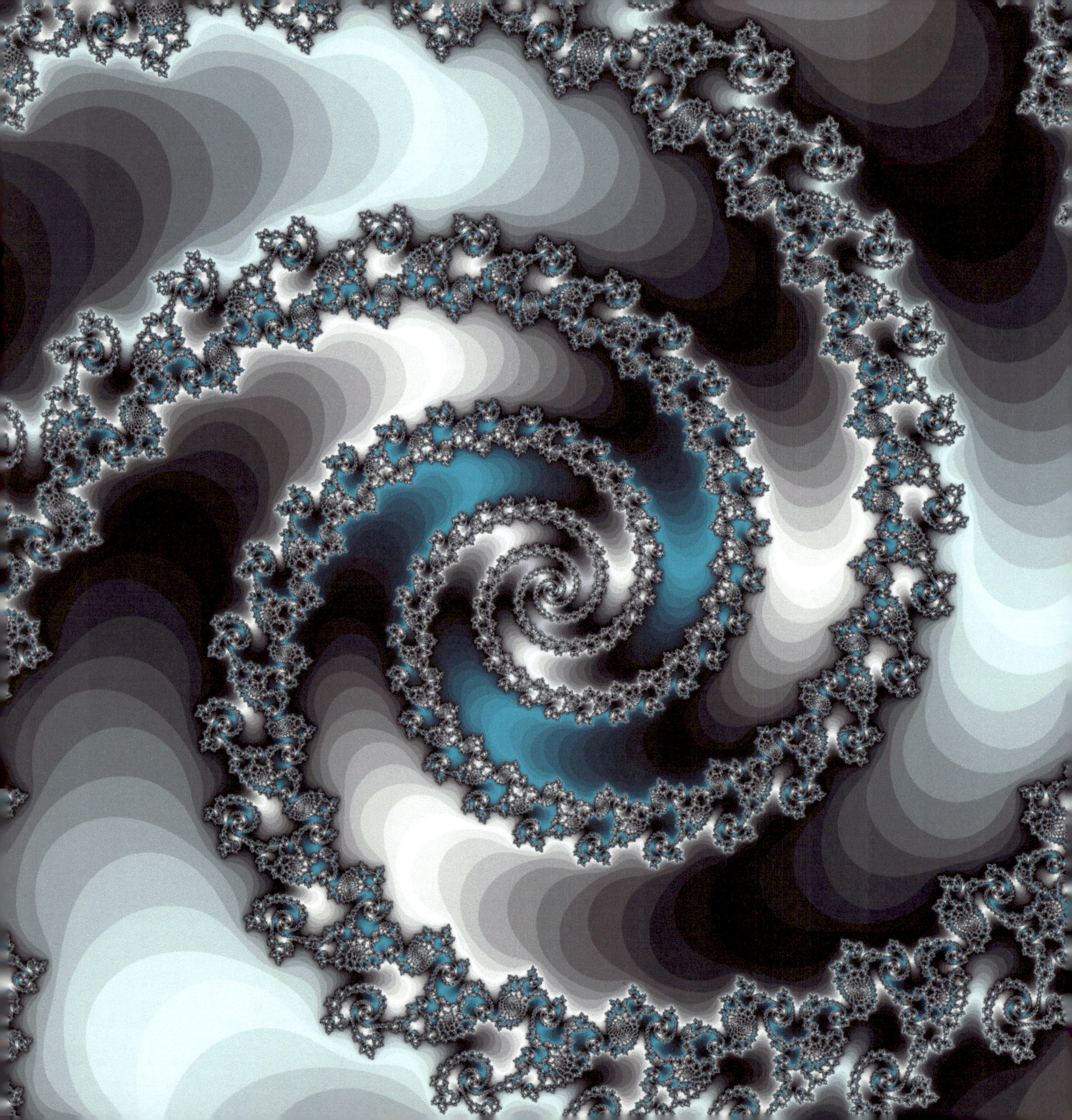

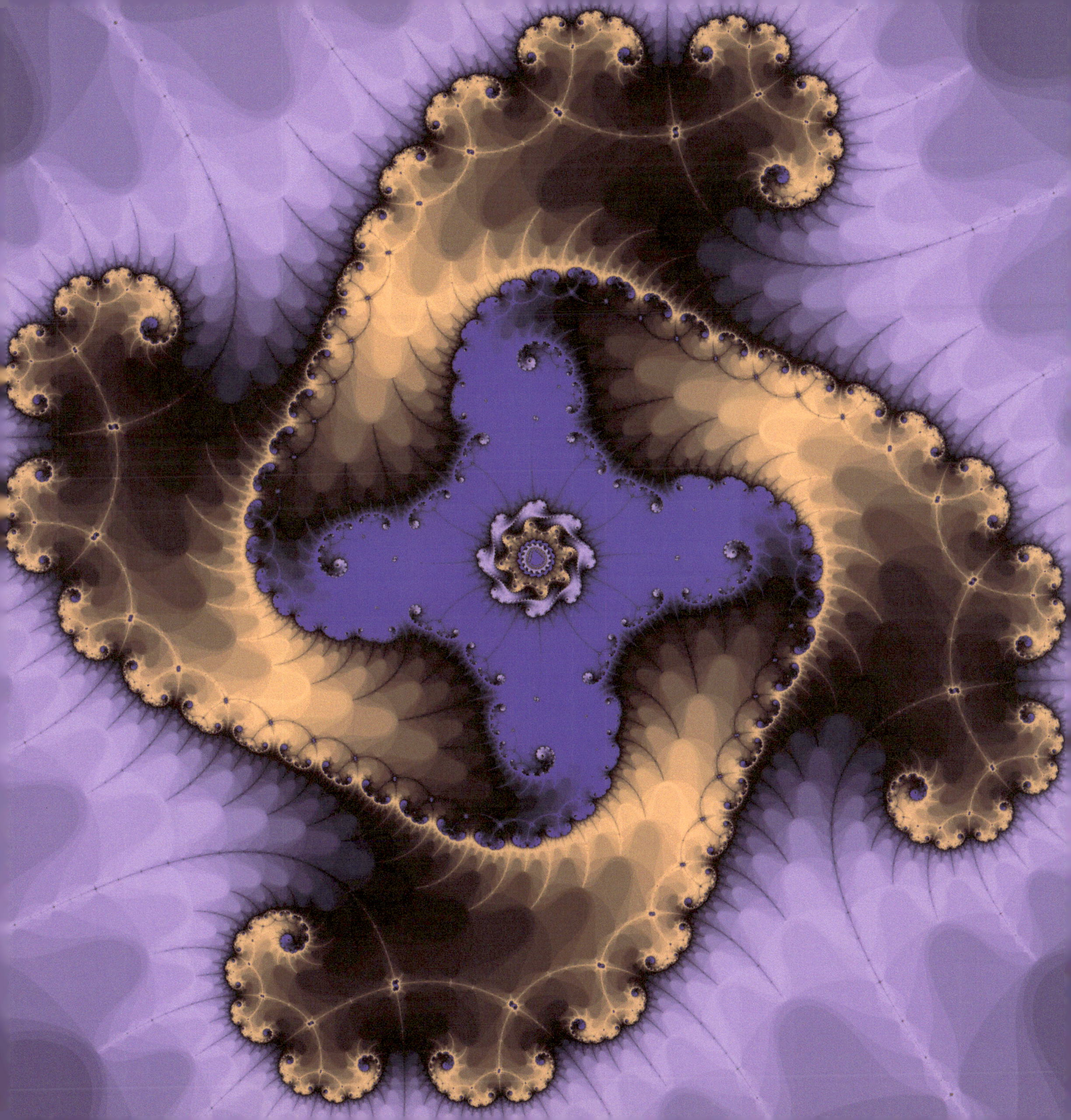

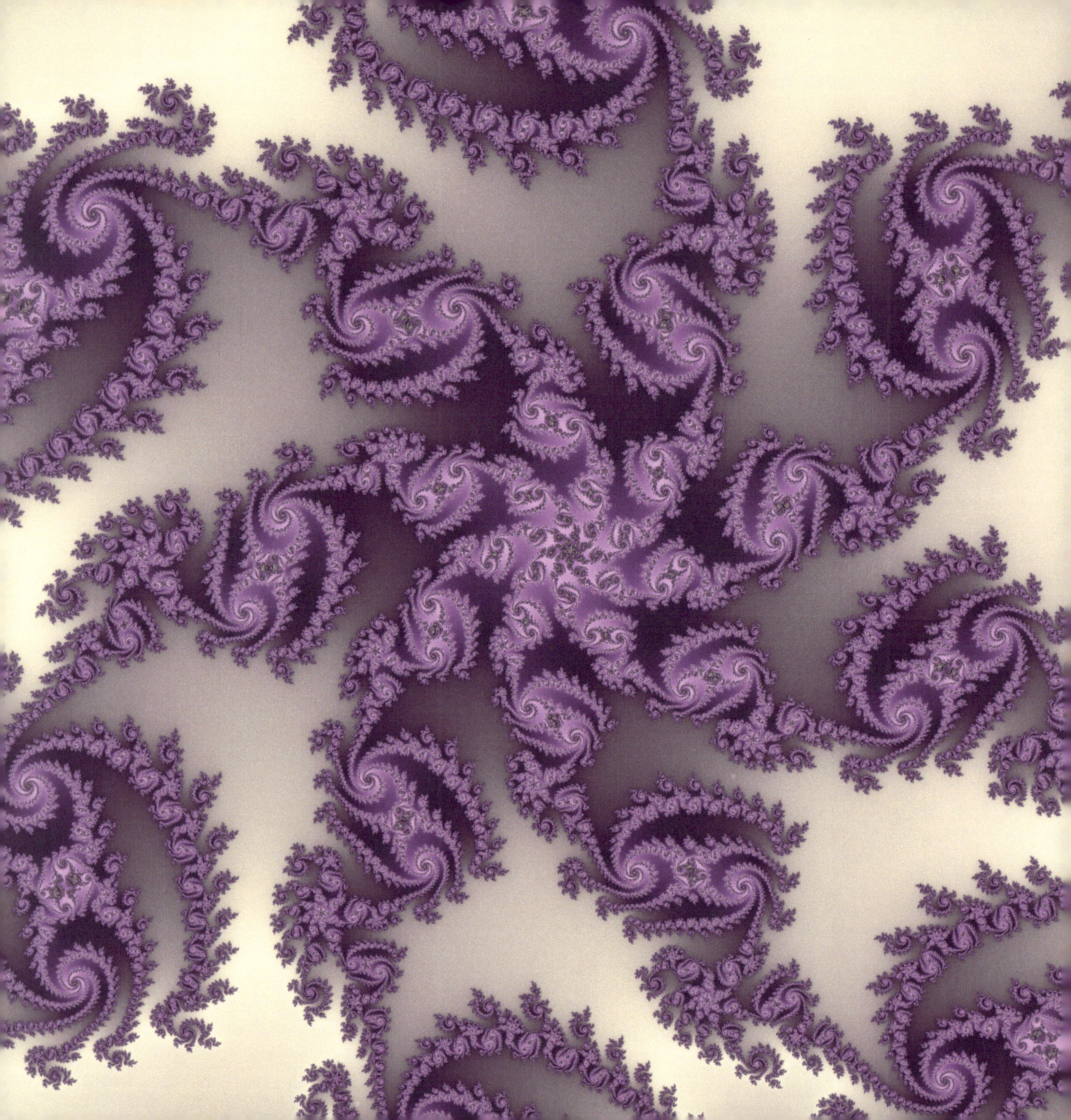

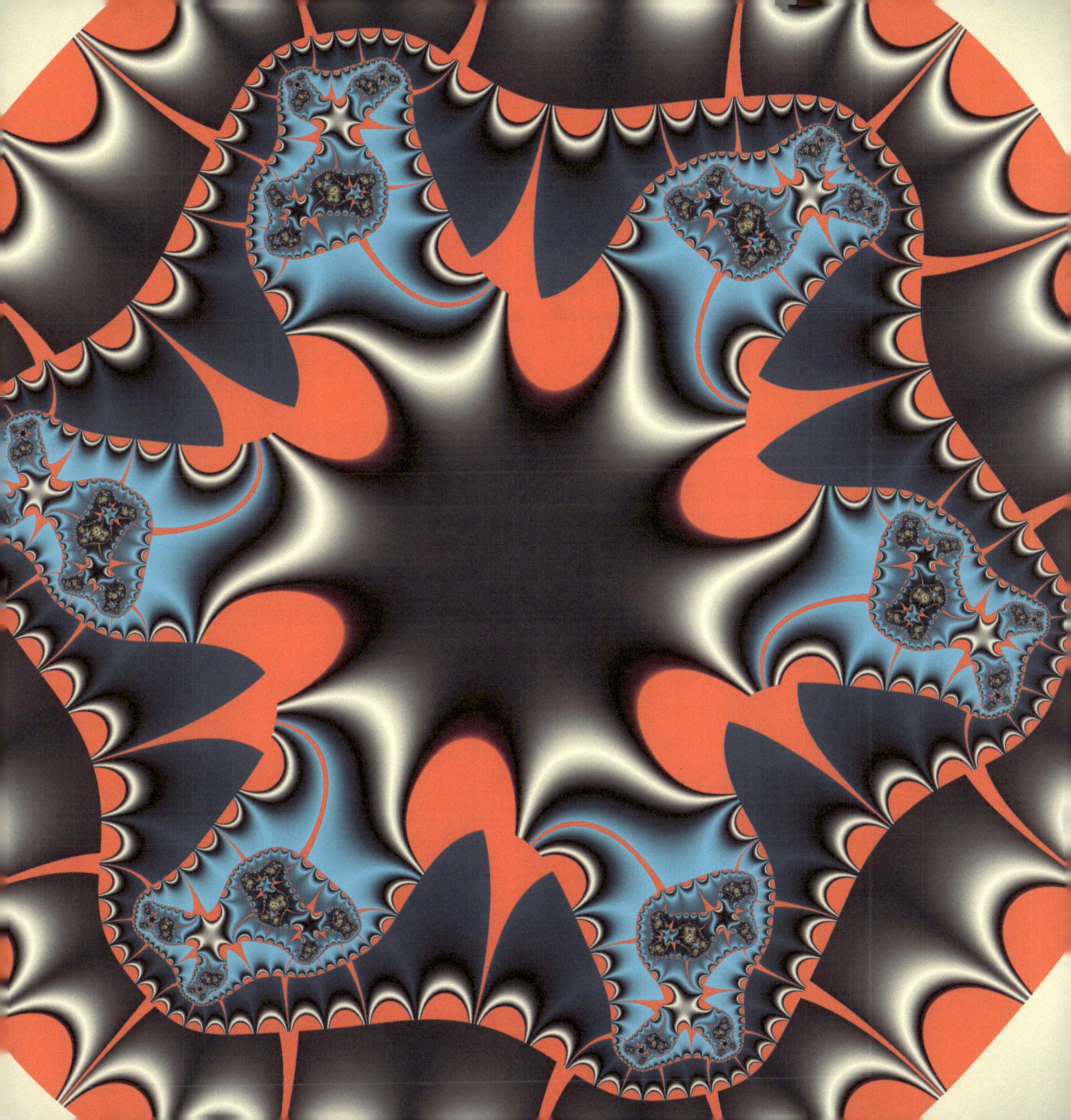

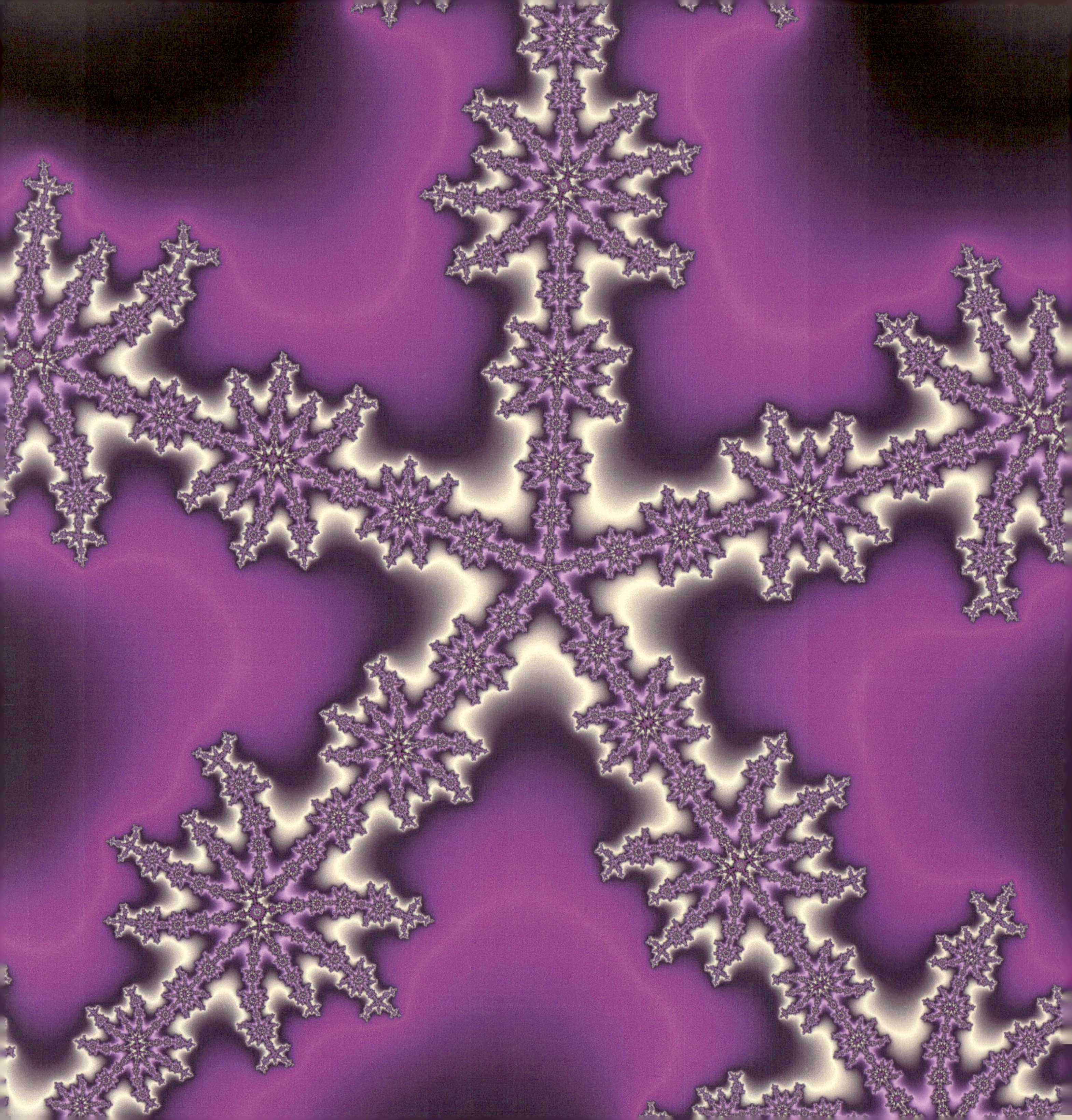

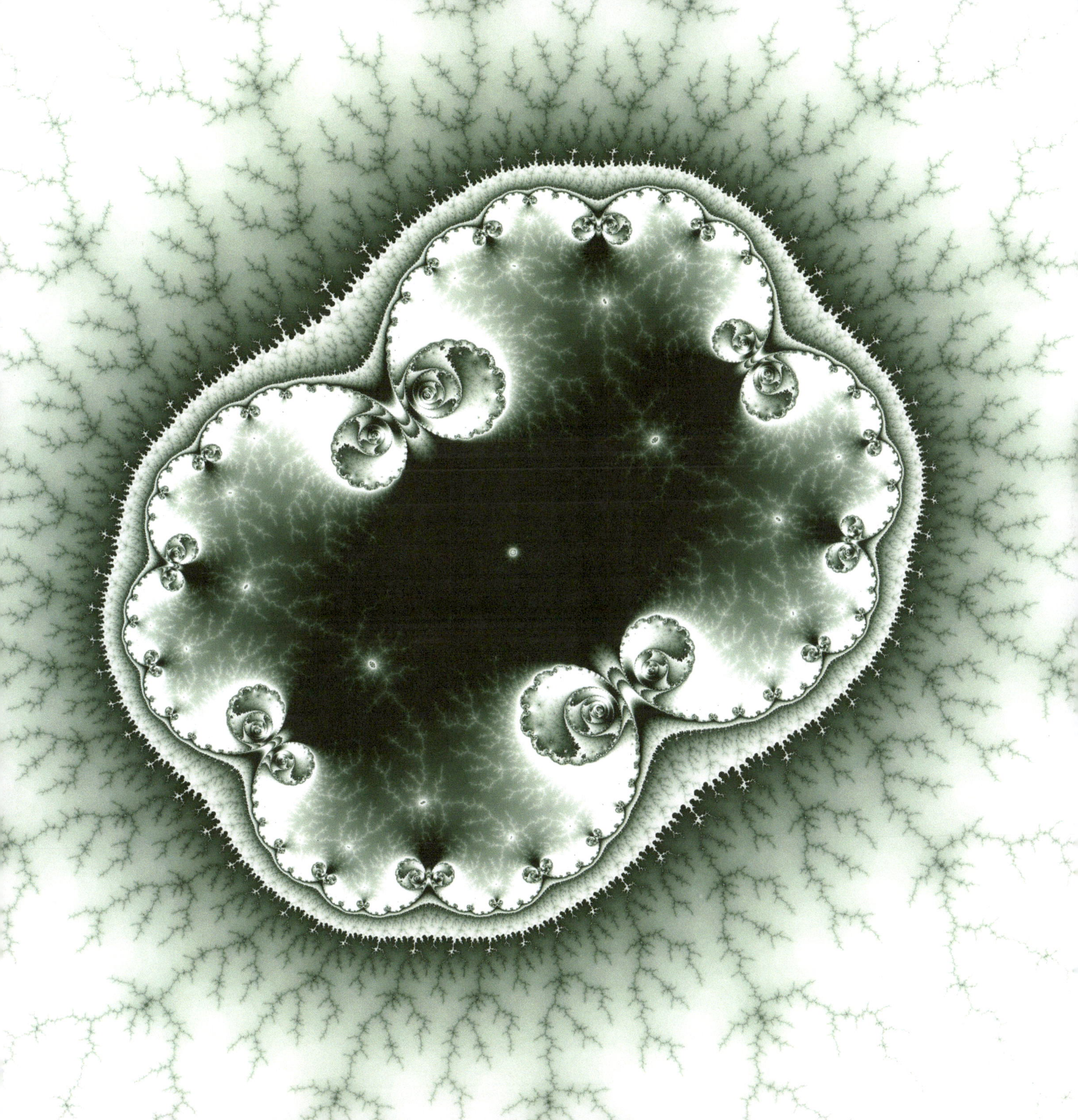

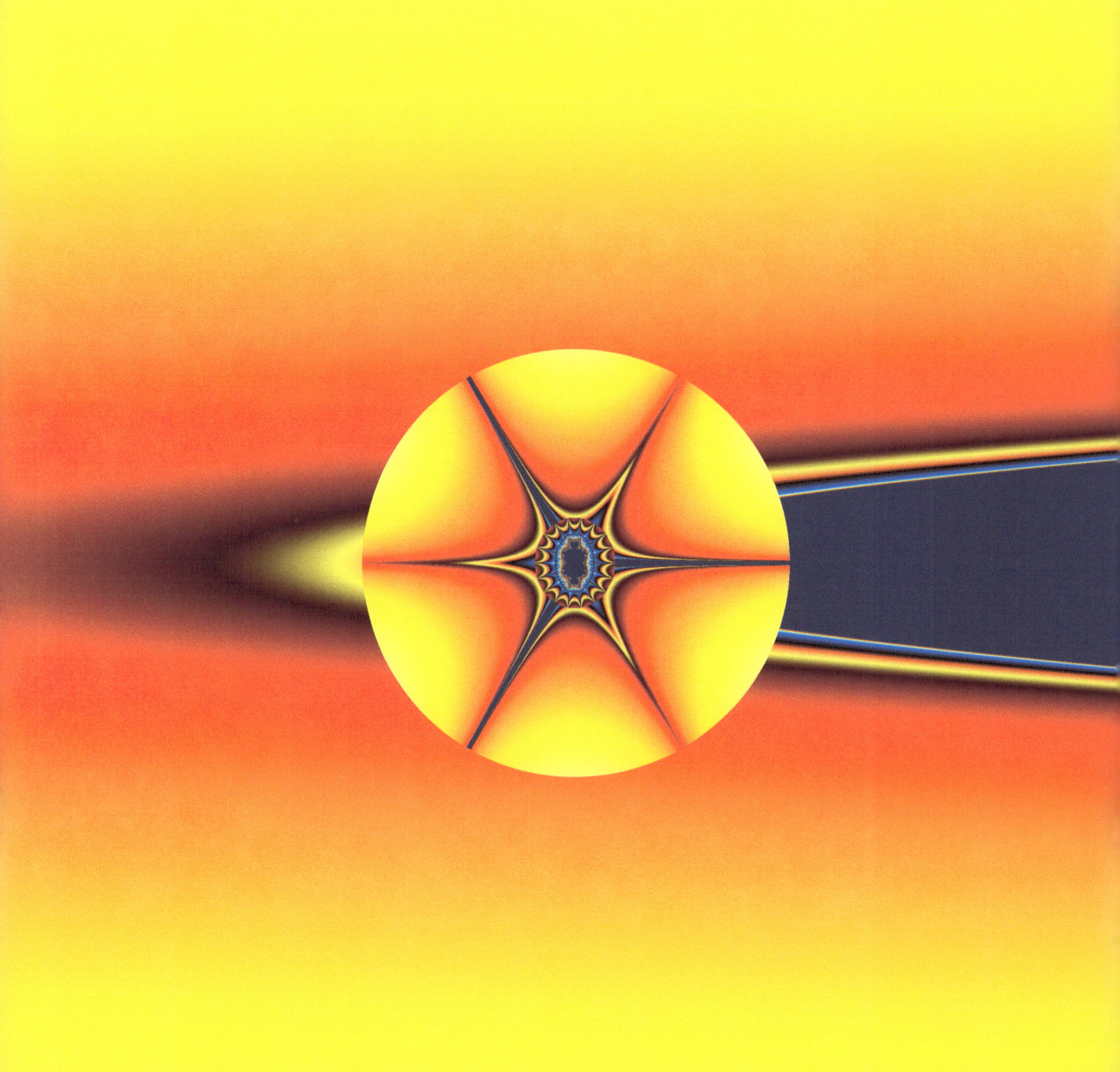

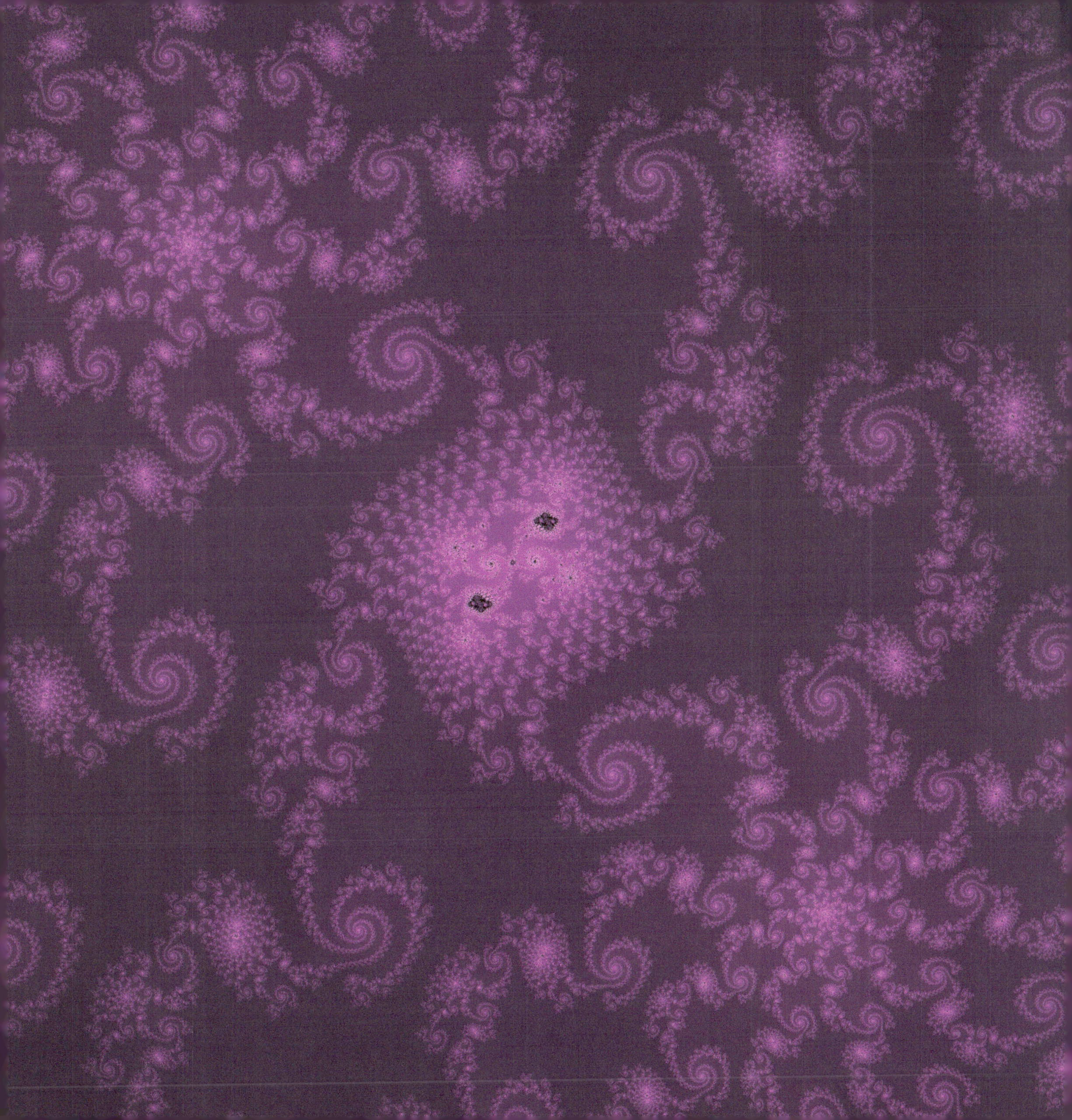

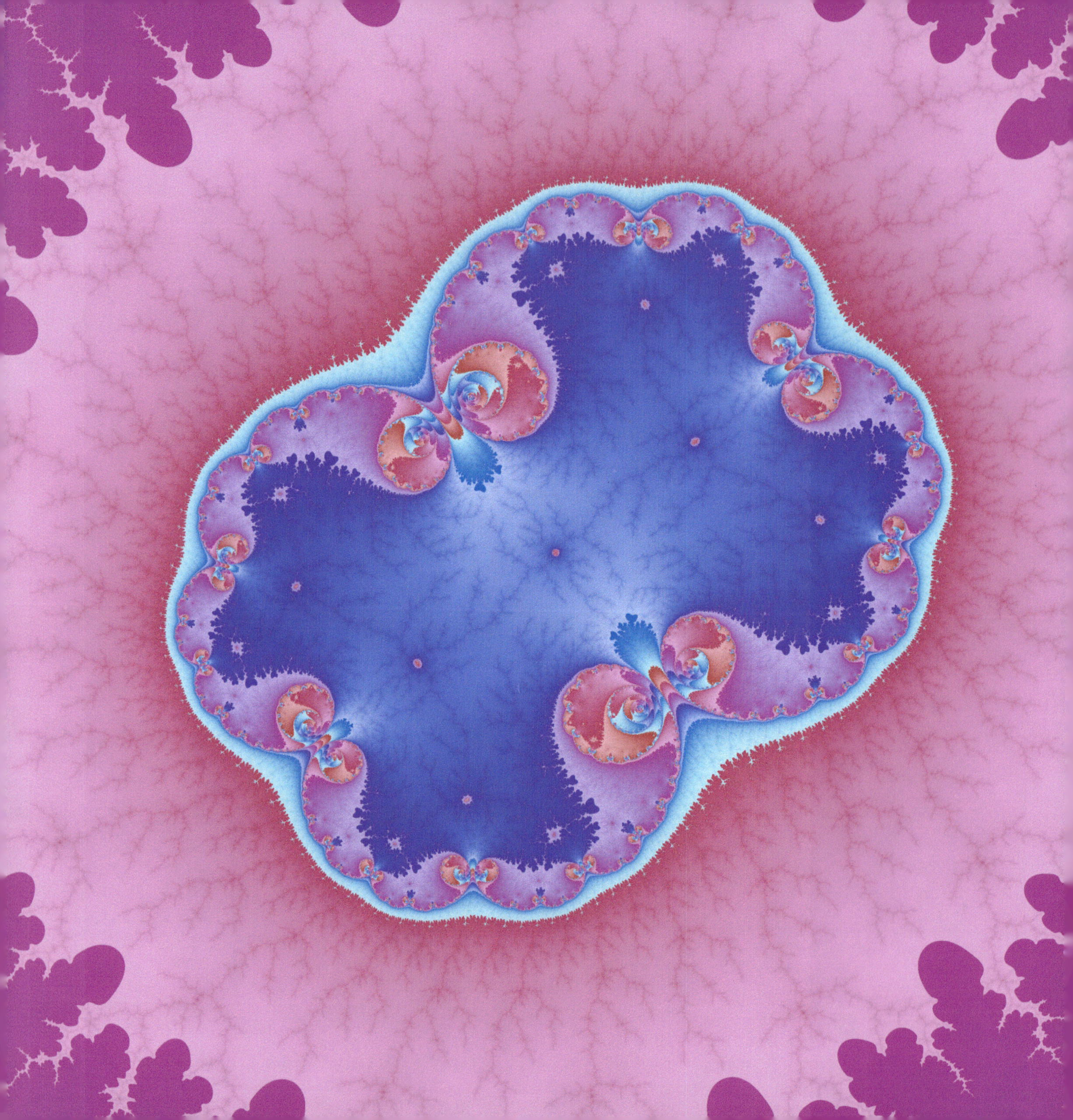